T0039125

CHITOSAN HYDROLYSIS
BY NON-SPECIFIC ENZYMES

CHITOSAN HYDROLYSIS BY NON-SPECIFIC ENZYMES

WENSHUI XIA

AND

PING LIU

Nova Science Publishers, Inc.

New York

Copyright © 2010 by Nova Science Publishers, Inc.

All rights reserved. No part of this book may be reproduced, stored in a retrieval system or transmitted in any form or by any means: electronic, electrostatic, magnetic, tape, mechanical photocopying, recording or otherwise without the written permission of the Publisher.

For permission to use material from this book please contact us:
Telephone 631-231-7269; Fax 631-231-8175
Web Site: http://www.novapublishers.com

NOTICE TO THE READER
The Publisher has taken reasonable care in the preparation of this book, but makes no expressed or implied warranty of any kind and assumes no responsibility for any errors or omissions. No liability is assumed for incidental or consequential damages in connection with or arising out of information contained in this book. The Publisher shall not be liable for any special, consequential, or exemplary damages resulting, in whole or in part, from the readers' use of, or reliance upon, this material. Any parts of this book based on government reports are so indicated and copyright is claimed for those parts to the extent applicable to compilations of such works.

Independent verification should be sought for any data, advice or recommendations contained in this book. In addition, no responsibility is assumed by the publisher for any injury and/or damage to persons or property arising from any methods, products, instructions, ideas or otherwise contained in this publication.

This publication is designed to provide accurate and authoritative information with regard to the subject matter covered herein. It is sold with the clear understanding that the Publisher is not engaged in rendering legal or any other professional services. If legal or any other expert assistance is required, the services of a competent person should be sought. FROM A DECLARATION OF PARTICIPANTS JOINTLY ADOPTED BY A COMMITTEE OF THE AMERICAN BAR ASSOCIATION AND A COMMITTEE OF PUBLISHERS.

LIBRARY OF CONGRESS CATALOGING-IN-PUBLICATION DATA
Xia, Wenshui.
  Chitosan hydrolysis by non-specific enzymes / Wenshui Xia, Ping Liu.
     p. cm.
  Includes index.
  ISBN 978-1-61668-152-4 (softcover)
  1. Chitosan--Metabolism. I. Liu, Ping. II. Title.
  QP702.C5X53 2009
  612.3'9--dc22
                                                            2009050570

Published by Nova Science Publishers, Inc. ✦ *New York*

CONTENTS

ABSTRACT

Chitosan, the only alkaline polysaccharide of β-1,4 linked N-acetyl-glucosamine and glucosamine,could be hydrolyzed by many non-specific enzymes such as cellulase, protease, and lipase, especially cellulase, which show high activity on chitosan. The hydrolytic mechanisms of these non-specific enzymes have been received growing attentions.

The focus of this chapter was the characterizations and hydrolyzing mechanism of the non-specific enzymes toward chitosan choosing the three typical non- specific enzymes: cellulase, lipase and papain as objects. We have studied the enzymatic characteristics, purification, product analysis, glycoside bond cleavage, active sites and gene cloning of these enzymes to expatiate their non-specific hydrolysis mechanism. From these, we obtained two bifunctional enzymes with chitosanolytic activity from commercial cellulase and lipase, respectively, and one chitosanase from papain.The three purified enzymes were the main reasons for the non-specific chitosanolytic hydrolysis of cellulase,lipase and papain, respectively. Moreover, It is identified that the bifunctional enzyme with chitosanolytic and cellulolytic activity(CCBE) from cellulase (*T.viride*) is identified as a cellobiohydrolase I with exo-β-D-glucosaminidase activity belonging to glycosyl hydrolysase 7 family.The enzmye with chitinase and chitosanase activity (CNBE)from lipase (*A. oryzae*) is the exo-β-D -glucosaminidase with N-acetyl-chitobiosidase activity belonging to glycosyl hydrolysase 18 family. Both of the two enzymes are novel and first reported in chitosanase families. Besides, the active sites and gene expression analysis of CCBE indicated that their dual activities originated from two distinct catalytic domains; while the two active sites overlapped partially.

Keywords: chitosan, cellulase, lipase, papain, chitosanase, non-specific enzyme, purification, characterization, gene cloning, expression

Chapter 1

INTRODUCTION

Chitosan, a linear polysaccharide composed of 2-acetamido-2-deoxy-β-D-glucose (N-acetylglucosamine, GlcNAc) and 2-amino-2-deoxy-β-D-glucose (D-glucosamine, GlcN) with various contents of these two monosaccharides, is a partial deacetylated derivative of chitin,which is the second most abundant polysaccharide on earth (after cellulose) and largely available in the exoskeletons of invertebrates and the cells of fungi [1]. Chitosan could be hydrolyzed enzymatically and chemically for the preparation of low molecular weight chitosans (LMWC) and chitooligosaccharides (COS). Enzymatic preparation methods captured a great interest due to safe and non-toxic concerns. Recently, chitosanolytic enzymes are increasingly gaining importance as LMWC and COS exhibit many outstanding biological properties, such as antimicrobial effect[2-8,18], antitumor and anticancer effect[2,7,9-11,18], antioxidant effect[2,7,13,14], immunostimulant effects[2,7,15-18], Antiatherosclerotic effects[18-20], Angiotensin I converting enzyme (ACE) inhibitory activity [21,22] and so on, which show innumerable applications in biomedical, pharmaceutical, agricultural, biotechnological and food fields[2,7,18].

In general, for preparation of COS and LMWC, chitosan could be degraded enzymatically by specific chitosanases and non-specific enzymes. The specific enzyme for chitosan hydrolysis has been found in a wide range of organisms, including bacteria [23, 24], fungi [25, 26] and plants[27]. Most of these chitosanases were characterized as endo-type and they split β-1,4 glycosidic linkages in chitosan in a random way to form chitosan oligosaccharides. But the utility of these specific chitosanases in hydrolysis is limited because of its cost and unavailability in bulk quantities. With regard to the non-specific hydrolysis on chitosan, in 1992, Pantaleone etc first found chitosan could be hydrolyzed by many kinds of enzymes such as cellulases,proteinases, pectinases, and lipases, which first raised the derivation of non-specificity[53].Since then, a number of non-specific commercial enzymes have been reported for their ability to degrade chitosan at a level comparable to that achieved by specific chitosanases, and have been used for preparation of chitooligosaccharides or low molecular weight chitosan from chitosan hydrolysis, such as cellulase[5,8,9,10,28-31], pectinase [32-36], pepsin[37], papain[38-42], amylase[43] and lipase[44-48], especially cellulases. The reactions are of interest as these commercial enzymes, especially food grade enzymes, have been used in food industries for decades and are safe and relatively inexpensive. Another advantage of non-specific chitosanolysis is the production of low molecular weight chitosan in higher yields due to their low specificity or nonspecificity [49].

However, few researches have been reported on the mechanism of these non-specific enzymatic processes, and even then controversial views exist. On the one hand, Kittur et al.[33] first reported that a multiple functional pectinase isoform from *Aspergillus niger* could be responsible for chitosan degradation by pectinases,but on the other hand, some authors have purified hetero chitosanases or/and chitinases from several commercial proteases,which were charged with their chitosanolytic activity [51-53].Whereas,no reports have been focused on non-specific hydrolysis of cellulases and lipases before, although their utility were popular in chitosan hydrolysis [5,8,9,10,28-31,43] and several bifunctional cellulase-chitosanases have been reported to be secreted from bacteria [58,79,87].

To make clear the non-specific mechanism for chitosan hydrolysis, taking the three typical non-specific enzymes: cellulase, lipase and papain as objects, we have been working on this research in detail and systematically for several years and have got a great progress. This chapter aims to the characterization and non-specific hydrolysis mechanism of these enzymes toward chitosan based on our research, which may provide a contribution for the development of chitin science.

Chapter 2

CHARACTERIZATION OF CHITOSAN HYDROLYSIS BY NONSPECIFIC ENZYMES

2.1. CHARACTERIZATION OF CHITOSAN HYDROLYSIS BY NON-SPECIFIC CELLULASES

Cellulase constitutes a complex enzymatic system responsible for the degradation of cellulose substances into glucose. It is obtained from bacteria and fungus, mainly from fungus such as *Trichoderma, Aspergillus, Penicillium*, and *Acremonium*. Especially, many commercial celulases are produced by *Trichoderma viride* and *T. reesei*. Cellulase is one of the two enzymes firstly applied in non-specific hydrolysis of chitosan [53, 54]. There have been many reports on the characteristics of chitosan hydrolysis by cellulases. Through investigating the optimum temperature and pH, pI, hydrolyzing kinetics, change of viscosity and distribution of products of cellulase on chitosan, we found that cellulases from different sources all show chitosanase activity, although their properties differ. In our previous research, the effects of temperature and pH, hydrolyzing kinetics, DD and metal ions on chitosan hydrolysis by a commercial cellulase from *T.viride* were studied in detail, the results showed that the optimal conditions of chitosan hydrolysis were at 60℃ and pH 5.2.The enzyme activity was increased by the addition of Na^+, Mg^{2+}, Ca^{2+}, Mn^{2+}, while it was strongly inhibited by the addition of Ag^{2+},Cu^{2+}, Hg^{2+} and glucose. It still retained 60.5% of its original activity at 60℃ for 1h.its best substrate was colloid chitosan with 90%DD. Meanwhile, product analysis indicated that this commercial cellulase hydrolyzes chitosan in both endo-split and exo-split manners, which could cleave the GlcN-GlcN bond as well as GlcNAc-GlcN bond [29].

Based on our study and compared to the reference, the characterizations of chitosan hydrolysis by non-specific cellulases reported were summarized as follows:

(1) Temperature and pH

The effects of temperature on an enzyme-catalyzed action are mainly embodied in two respects: one is the increase of the reaction rate because a higher temperature accelerates molecular collisions between the enzyme and the substrate; the other is the inactivation of the enzyme for a higher temperature denatures the enzyme. The optimum temperature of cellulase acting on chitosan was in the ranges of 30-70 ℃. The optimal temperature of cellulase secreted by fungus on chitosan was between 50-60 ℃, higher than that from bacteria, which was often lower than 40 ℃.

The pH affects not only the dissociation behavior of the substrate but also the space structure and dissociation state of the active group in the enzyme, therefore, pH is among the most significant factors affecting the enzyme-catalyzed reaction. The variations in optimum pH of its chitosanolytic activity between the sources were not obvious. Most of the optima were focused in the range 5.0-7.0, few in particular: the optimum pH for a commercial cellulasewere 4.0 [31], while *Streptomyces griseus* cellulase acting on chitosan had optimum pH at 8.0 [55].

(2) Molecular Weight

Most of the cellulases that show chitosanolytic activity have the apparent molecular mass in the range of 23-55kDa, which is consistent with that of specific chitosanases. But there are also few exceptions such as: the molecular weights of those from *T. viride* [29, 56] and *T. reesei* PC-3-7[57] are 66kDa and 97kDa, respectively.

(3) Reduction of the Viscosity

Almost all the cellulases from different sources can decrease the viscosity of chitosan extensively in a short time; more than 60% of viscosity was reduced in the early stage (30 min or 1h) of the reaction, which suggestes that the cellulases possess the chitosanolytic activities of the endo-type manner. Pantaleone et al found three cellulases derived from *A. niger*, *T. viride*, and *T.reesei* enable to result in a 99% reduction of the viscosity [53, 54]. Several researchers at home also found the similar characteristics [8-10, 28-31].

The distribution of the reaction products confirms that most of cellulases from various sources show an endo-acting nature, they predominantly liberate a dp2-10 mixture, made up of dimers, trimers and oligomers from chitosan, though there are few in exo-type, for instance, the bifunctional enzyme purified from a commercial cellulase [29,56] and from *T.reesei*[57]showed an exo-β-D-glucosaminidase activity on chitosan.

(4) Kinetic Parameters

The Km and Vmax values reported for cellulases on chitosan differ from organism to organism and from those on cellulase [56] which means such chitosanolytic cellulases have different affinity with chitosan and cellulose ----two substrates with high similarity in structures: for Km value is a index reflecting the affinity of enzyme and substrate. From this, we assume that the cellulases with dual activities probably have two substrate binding domains or different binding sites were involved in a single binding domain for chitosan and cellulose, which needs further research.

(5) Metal Ions

Generally speaking, metal ions are not involved in the catalytic activities of chitosanases and cellulases[43,55,56,58,59], but they activate or inhibit their activity. The activities of these chitosanases and cellulases could be inhibited by heavy metal ions Ag^+, Hg^{2+}, Cu^{2+}, which form coordinate bonds with the side chain group of the enzyme, resulting in the change of protein conformation hence inactivating the enzyme. While Mg^{2+}, Mn^{2+} etc could enhance the molecular conformational stability, accelerating the catalysis[60], hence, enzymes incubated with Mg^{2+} etc generally show higher catalytic activity compared to the native (measuring under the same condition). The chitosanase activity of cellulases were inhibited strongly by Ag^+, Hg^{2+} etc heavy metals, and stimulated by Mg^{2+}, Mn^{2+}, which is similar to its cellulolytic activity[29,56,58].These phenomenon indicates that the active sites for the two activities of cellulase have some similarities.

2.2. CHARACTERIZATION OF CHITOSAN HYDROLYSIS BY NON-SPECIFIC LIPASE

Lipases (EC3.1.1.3) are glycerol ester hydrolases, they can catalyze the hydrolysis of triacyglycerols into free fatty acids, partial acylglycerols and glycerol [63].Lipase was found to exert chitosanolytic activity. Although Pantaleone et al first found that the lipase from two *A. niger* and many glucanases showed highest non-specific chitosanolytic activity,much more than that of protease[53], there are still few reports on chitosanolysis of lipases up to now [44, 61, 62]. Muzzarelli et al firstly took a detail research about the chitosan hydrolysis by lipase. They found that the wheat germ lipase preparation was very active in depolymerizing chitosan and modified chitosans dissolved in slightly acidic aqueous solutions. This enzyme preparation was effective in drastically lowering the viscosity of chitosan, MP-chitosan, and NCM-chitosan solutions within a few minutes. Typically, the viscosity was lowered to 35% of the initial value upon 10-min contact with lipase at 25°C. Measurements taken in the lipase concentration range 4.5 mg/L -0.9 g/L showed a logarithmic dependence of the initial velocity on the enzyme concentration. N-Carboxymethyl chitosan was an even better substrate, with the initial velocity over double that for chitosan [44]. Since then,in 2001,Shin et al studied the degradation of chitosan with the aid of lipase from *Rhizopus japonicus* for the production of soluble chitosan,finding that the lipase could degrade chitosan to water-soluble chitosan with Mw between 30 kDa ~ 50 kDa at its optimal temperature of 40°C[62]. Many researchers at home also found that lipase could degrade chitosan effectively, the reaction did not follow the classic Michaelis-menten equation,and the optimum temperature were in a range of 40°C~55°C,while the optimal pH differ significantly, such as pH 5[64],pH3 [46] and pH5.0~5.5[47].However, there has been no report on qualitative analysis of chitosan oligomers resulted from the chitosan hydrolysis aided by lipase.

In our recent study, we investigated the effects of a commercial lipase from *A.oryzae* on chitosan hydrolysis with different conditions such as pH,temperature,degree of deacetylation (DD) and molecular weights, metal ions,viscosity reduction, and qualitatively analyzed COS products using kinetic analysis,thin layer chromatography (TLC) and HPLC methods[48].

The optimal temperature and pH of the lipase was 55°Cfor all chitosans. Considering that DD and distribution of N-acetyl groups significantly affected the properties of chitosan solution [65], in our research, it was found that when four chitosans with various degrees of deacetylation were used as substrates, the lipase showed higher optimal pH toward chitosan with higher DD,among pH4.5-4.8.The enzyme exhibited highest activity to chitosans which were 82.8% and 73.2%

deacetylated, indicating that the presence of GlcNAc residues in the chitosan molecules is important for the enzyme to exhibit chitosanolytic activity and the enzyme recognized not only the GlcN residues, but also the GlcNAc residues in the substrate. In addition, it can be seen that when chitosans with the same DD were used as the substrates, chitosans with lower molecular mass were more susceptible to be hydrolyzed by the enzyme than those withhigher DP (degree of polymerization).Kinetics experiments show that chitosans with DD of 82.8% and 73.2% which resulted in lower Km values(3.588mg/mL and 3.754mg/mL, respectively) had stronger affinity for the lipase and the lipase action on all chitosans obeyed the Michaelis-Menten kinetics (data not shown).

Time course analysis of chitosan hydrolysis with the aid of lipase exhibited that there was a sharp decrease in viscosity of the mixture during the early hydrolysis stage at both 37°Cand 55°C temperatures, although there was a very weakly marked difference in viscosity;besides,the chitosan hydrolysis carried out at 37°Cproduced larger quantity of COS (chitooligosaccharides) than that at 55°C when the reaction time was longer than 6 h, and COS yield of 24 h hydrolysis at 37 °Cwas 93.8%. Hydrolysis of chitosans with different DD produced the same series of products,viz., glucosamine and chitosan oligomers with DP from 2 to 6 and above. This means that chitosans with different degrees of deacetylation did not result in different products.and the lipase acted on chitosan in both exo- and endo-hydrolytic manner. According the dynamics of COS and chitosan hydrolysis, it can be seen as follows: at the early stage of hydrolysis, the lipase cleaved glycosidic bonds in both endo- and exo-mode, and GlcN was first produced, and COS with higher DP were then produced. During this time, viscosity of the mixture decreased dramatically. Then the lipase hydrolyzed the oligomers mainly in an exo-mode, therefore, the oligomers gradually disappeared, the amount of GlcN increased, and the viscosity of the action mixture almost remained the same. Moreover, the FT-IR analysis of the chitosan hydrolysates suggested that the lipase had loose substrate specificity, which didn't change the side-chain conformation and degree of deacetylation of the products during chitosan hydrolysis,which is not similar with that from papain[49].This result is useful in the direction of production of COS. If the objective is to obtain oligomers with higher DP, the reaction time should be strictly controlled in order to avoid the production of large quantity of monomers; on the other hand, this lipase can be used to prepare the chitosan monomer, glucosamine, which is widely used to relieve arthritic complaints [66].

2.3. Characterization of Chitosan Hydrolysis by
Non-Specific Papain

Many proteases such as papain [38-42,53], pepsin[37], trypsin[67], bromelain[53,68], ficin[50,53] etc ,all can degrade chitosan,among these,papain was found to depolymerize chitosan efficiently [53]. It's been reported that papain degraded chitosan under a optimal conditions of pH 3.0-4.0 and temperature of 40-50 °C ,although different papains were used in different research, and the high molecular weight of chitosan with a Mw of $4-5\times10^5$ were degraded by papain to produce the LMWC and COS in a range of 10^3-10^4.

Muzzarelli et al [69] studied chitosan deploymerization by papain in the form of its lactate salt at acidic pH values by papain. The results showed that this enzymatic process did not obey a simple kinetic model; the initial velocity was, however, strongly enhanced by high substrate concentration, while the temperature had little influence. A viscosity decrease as high as 94% could be obtained in 1 h with free papain at pH 3.2 and 50°C. Initial velocity for a $19g.L^{-1}$ solution at 25°C and pH 3.2 was 110 mPa .s . min^{-1}. Chitosans with average molecular weights in the range $4-7 \times 10^5$ could be easily deploymerized to highly polydisperse chitosans. Modified chitosans were also depolymerized, though with lower initial velocities. Once immobilized on chitin, papain could be used repeatedly to deploymerize chitosan lactate salt with no observable loss of activity. The data are of value for the production of chitooligomers of medical and biotechnological interest.While Yalpani[54] found that at lower concentration(0. 5% -1%) , the concentration of substrate did not affect the chitosan hydrolysis by papain and papain could be repeatedly used at least 10 times when immobilized on chitin or chitosan with no evident lost of its chitosanolytic activity.Under the optimum conditions: NaAc–HAc buffer system, pH 4.0, temperature 45 °C, immobilized papain was used to prepare chitosan oligomers, After 24 h reaction with stirring, the yield of chitosan oligomers with molecular weight under 10,000 was 49.55% to total substrate, while that of between 600 and 2000 was 11.07%. The results showed immobilized papain could depolymerize chitosan successfully, and separating the product by ultrafiltration membrane was feasible. An amount of 1mg free papain and immobilized papain could produce 2.57 and 17.34 mg chitosan oligomers, respectively. The results indicated that the depolymerization efficiency of immobilized papain was obviously higher than that of free papain[70].

Terbojevich et al[42]studied the molecular parameters of chitosan depolymerize with the aid of papain,finding that the chitosan with the highest degree of polymerization were preferred and its Mw,viscosity and RG values were all lowered. The papain mainly cleaved the GlcN-GlcNAc bond in chitosan. However, Su et al reported that papain could split the GlcNAc-GlcN as well as

GlcN-GlcN bond and produced the mixture of LMWC and COS with DP of 2-6[38]. Kumar found the similar results using papain from Papaya latex and protease from *Streptomyces griseus* under each optimal conditions [49],Low molecular weight chitosans (LMWC) of different molecular weight (4.1–5.6kDa), which were obtained by the depolymerization of chitosan using papain ,could be tested by Scanning electron microscopy (SEM) ,circular dichroism (CD), FT-IR and solid-state CP-MAS 13C-NMR data,while LMWC with different Mw could be isolated by ultrafiltration and Sephadex G-50 gel filtration.

Chapter 3

MECHANISM OF NON-SPECIFIC ENZYMES TOWARD CHITOSAN

Considering there are many reports on the characterization of chitosan degradation by these non-specific enzymes,while their mechanisms are still controversial,we studied the mechanism of these non-specific chitosan hydrolysis taking the typical cellulase,lipase and papain as objects.Using two purification systems: one is column chromatography combining with ultrafiltration, ion exchange, hydrophobic interaction chromatography and gel filtration and the other is gel electrophoresis,we have found that the non-specific hydrolysis of these enzymes toward chitosan were due to the existence of the bifunctional enzymes with chitosanolytic activity in cellulase and lipase[56,71] or hetero chitosanases in papain[72] (seen in Table1). Bifunctional enzymes with chitosanolytic activity was focused much attention since its novelty and contravention to the traditional enzymological theory----one enzyme shared one or one kind of catalytic activity.

With regard to the conception of multiple-functions, it has been presented for about two decade's years. Multiple substrate specificities of a number of glycanases and glycosidases have been known in literature. β-1,4-Glucanases with hydrolyzing activity on mannan has been reported from *T.reesei*[73]. β-Glycosidases (designated as BgIA and BgIB) from *Bacillus* sp. GLI have been shown to cleave α- and β-linkages in p-nitrophenyl (pNP)-glycosides and positional isomers of β-1,4-glycopyranosyl linkages [74]. The product of BgIB gene was also identified as a gellan degrading enzyme. Pectinase with chitosanolytic activity have been reported from *A.niger* [33]. Some bifunctional enzymes with chitosanolytic and cellulolytic activity have been also reported from the Prokaryotes [58, 75-83].

Table 1.The parameters of the purified chitosanolytic components from commercial enzymes

parameters	CCBE	PCh	CNBE
source	cellulase/*T.viride*	papain	lipase/*A. oryzae*
Purification methods[a]	DEAE-Sepharose CL-6B ion-exchange chromatography, Phenyl Sepharose CL-4B hydrophobic interaction chromatography, Sephadex G-75 gel filtration	Ultrofiltration and Gel electrophoresis	DEAE-Sepharose CL-6B ion-exchange chromatography, Phenyl Sepharose CL-4B hydrophobic interaction chromatography, Sephacryl S-200 gel filtration
Mw/kDa	66	31	130
subunit	single	single	Two identical subunits/ 71kDa
pI	4.55	----	---

a :the same purification system used for CCBE and CNBE,but each had its own eluting conditions.

Several explanations have been proposed for such enzyme actions on completely unrelated substrates. Gradually the concept of one enzyme-one activity is vanishing and several examples of multifunctional enzymes are being identified.The latter may result from gene sharing, gene fusion andexon shuffling. This segment describes the non-specific hydrolysis mechanism of chitosan by a commercial cellulase and lipase, since both of them contained bifunctional chitosanolytic enzymes mainly responsible for its chitosan hydrolysis.

3.1. PURIFICATION AND CHARACTERIZATION OF CHITOSANOLYTIC COMPONENTS FROM NON –SPECIFIC ENZYMES

3.1.1. Purification and Characterization of a Bifunctional Enzyme with Chitosanolytic and Cellulolytic Activity from Commercial Cellulase

Most cellulases show a non-specific hydrolytic action on chitosan, and some are even superior to specific chitosanases in degrading chitosan. This may be derived from the structural similarity of chitin, chitosan and cellulose, which are all polymers of D-glucose linked by β-1,4-glycosidic linkages (in chitosan, the C-2 hydroxyl groups are replaced by amino groups), and it seems that the enzyme does not recognized so severely the group of C2 position in glucose or glucosamine residue when enzyme-substrate complex was formed. For the same reasons, the chitinases, chitosanases and cellulases also exhibit high homogeneity and often appear in the same microorganism.

More than one bifuntional enzymes with chitosanolytic and cellulolytic activity have been reported in literature. An enzyme that has both β-1, 4-glucanase and chitosanase activities from *Myxobacter sp.* A-L1 was first purified and characterized by Hedges and Wolfe in 1974 using citrate extraction, ammonium sulfate precipitation, SP-Sephadex elution and G-75 elution, meanwhile evidence for homogeneity was obtained from electrophoresis and sedimentation velocity studies and presence of one N-terminal amino acid, valine. This enzyme caused a shift in attention to the non-specific chitosanase activity of cellulase [58]. Since then, there were many other bifunctional enzymes with chitosanolytic and cellulolytic activity purified from *Streptomyces griseus* [76,77], *Bacillus. megaterium* P1[78], *B. circulans* WL-12[79], *Bacillus sp.* 7-M[80],*B. cereus* S1[81], *Bacillus sp.* 0377BP [82], *B. licheniformis* NBL420[83], *B. cereus* D-11[84], *B. cereus* D-11[85], *B. cereus* S-65[86] and *Lysobacter sp* IB-9374[87] using the conventional protein purification techniques such as ammonium sulfate fractionation, gel filtration, ion-exchange chromatography, affinity chromatography, hydrophobic interaction and isoelectric focusing,although their nominations based on its activities were not uniform:some called as cellulases with chitosanase activity[79,87],some were chitosanases with carboxylcellulose degrading activity[76-78,80-86],while some were named as bifunctional cellulase-chitosanases [58,75]. Moreover, except two from *Streptomyces griseus*, all of the above bifunctional components were produced from bacteria, suggesting that the bifunctional cellulase-chitosanases were responsible for chitosan non-specific hydrolysis of bacterial cellulases.

In industry, cellulases were mainly produced by fungi. However, although many commercial cellulases have been reported to be used for preparation of chitooligosaccharides or low molecular weight chitosan, and the commercial cellulases are generally produced from fungus such as *T.viride* and *T. reesei* under the induction of cellulose, there are almost no reports on non-specific mechanism of chitosan degradation by fungal cellulases.

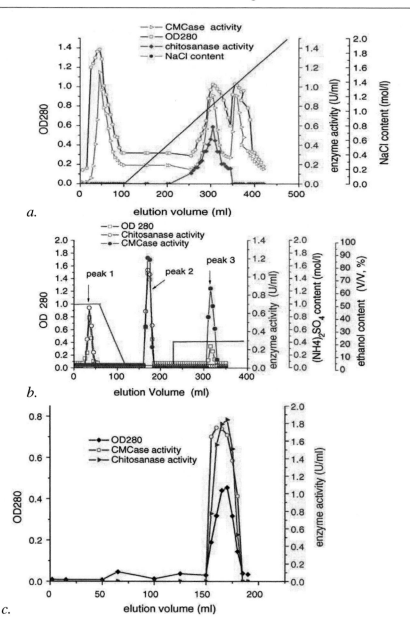

Figure 1.Purification of the bifunctional enzyme with chitosonalytic and cellulolytic activity from commercial cellulase with column chromatography. (a). Ion-exchange chromatography of cellulase on DEAE-Sepharose CL-6Bcolumn;(b).Hydrophilic chromatography of cellulase on Phenyl Sepharose CL-4B;(c).Gel filtration of cellulase on Sephadex G-75 column.(Cited from reference [56]).

In this chapter, the chitosan degrading mechanism of fungal cellulases was studied [56,88]. Through sequential steps of DEAE-Sepharose CL-6B ion-exchange chromatography, Phenyl Sepharose CL-4B hydrophobic interaction chromatography and Sephadxe G-75 gel filtration, a bifunctional enzyme with chitosanase and carboxymethyl cellulase (CMCase) activity (CCBE) was first purified from commercial cellulase produced by *T.viride*. As shown in Fig1,when eluted from the ion-exchange chromatography of DEAE Sepharose CL-6B, a protein peak containing chitosanase and CMCase activities and two other protein peaks just with CMCase activity were eluted(Fig1a).Fraction with bifunction was collected, and others were removed using this procedure. Hydrophilic chromatography on Phenyl Sepharose CL-4B resulted in one bigger protein peak (peak2) with bifunction and two smaller protein peaks just with chitosanase (peak 1) or cellulase activity (peak 3)(Fig1b).From this, it can be seen that more than one chitosanolytic components existed in the commercial cellulase, while the protein of peak 2 were the main chitosanolytic component. So enzymes obtained from protein 2 of Phenyl Sepharose CL-4B were further purified on a Sephadex-G-75 column and a solo protein peak was obtained as shown in Fig1c. Through these steps, the enzyme was purified 3.16-fold of its chitosanase activity with a recovery of 10.47%, while its purification fold of its CMCase activity was 1.05,not in agreement with that of chitosanase, maybe owing to three factors: the degrading CMC is a synergistic action of coenzyme but most of the cellulase active portion was removed during the purification procedure; the CMCase activity was determined by 3,5-dinitrosalicylic acid method which only measuring the reducing sugar; the bifunctional enzyme may has two different active centers, although one enzyme possesses both activities.

The purified bifunctional hydrolase (CCBE) was identified homogeneous by three methods of sodium dodecyl sulfate-polyacrylamide gel electrophoresis (SDS-PAGE), RT-HPLC and isoelectric focusing. The molecular mass of CCBE was 66kDa as estimated by SDS-PAGE(Fig2d), which was a little higher than those of the counterparts from bacteria, the latter have the apparent molecular mass in the range of 23-55 kDa. IEF showed that CCBE's pI was about 4.55(Fig2f).Activity analysis showed that the CCBE exhibited comparable chitosanase activity and CMCase activity, as well as those purified from *B. cereus* D-11[85], *Myxobacter sp.* AL-1[58,75] and *B. circulans* WL-12[79]. Considering that Cel8A from *Lysobacter sp.* IB-9374[88] exhibits a chitosanase activity that is 15–40% of its cellulase activity; while for *B. cereus* S1[81] and *Bacillus sp.* 0377BP[82] bifunctional chitosanases, their CMCase activity was equal to 3–30% of its chitosanase activity,we can infer that the dual activities of the bifunctional enzymes differ with sources, but have no rules.

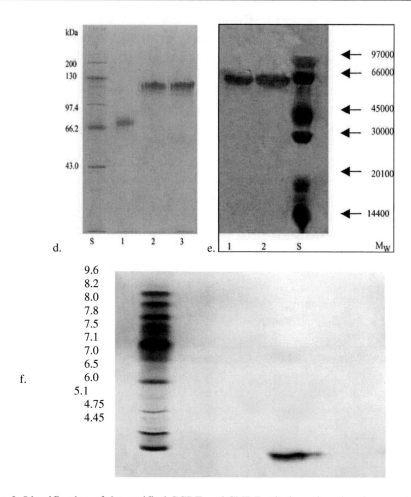

Figure 2. Identification of the purified CCBE and CNBE. (d).the reduced and non-reduced SDS-PAGE analysis of the purified CNBE.lane1:reduced analysis;lane 2and 3:non-reduced analysis.Lane S indicated moderate molecular weight protein standard;(e). SDS-PAGE of purified CCBE. Lane 1 and 2: purified CCBE;Lane S indicated low molecular weight protein standard.SDS–PAGE was performed in 10% polyacrylamide gel and visualized by Coomassie Brilliant blue(d) and silver staining(e);(f).IEF analysis of the purified CCBE. (Cited from reference [56,88]).

The CCBE showed different optimum condition and kinetic parameters on the two substrate: chitosan and CMC (seen in Table 2).The optimum condition of CCBE on chitosan hydrolysis were pH 5.2 and 55℃, Km and V max were 0. 746 glucosamine mg/mL and 0.0268 mg.mL^{-1}.min^{-1}, respectively. Its chitosanase activity was stable at pH 4.0-7.8 and 30-50℃.It was gradually inactivated at 60℃

and significantly lost activity at 80℃. However, the optimal condition toward CMC were pH 4.2 and 55℃, Km and Vmax were 4.087 glucose mg/mL and 0.127 mg.mL^{-1}.min^{-1}. Its CMCase activity was stable at pH4.0-6.0 and its temperature stability was consistent with that toward chitosan.From this, it can be seen that the CCBE probably had two distinct catalytic domains toward chitosan and CMC.

Table 2. The characterization of purified CCBE and CNBE

parameters	CCBE		CNBE
	CMC	chitosan	chitosan
Optimal pH/pH stability	4.2/4.0-6.0	5.5/4.0-7.8	4.8/4.5-9.5
Optimum temperature/stability /℃	55/30-50	60/30-50	60/40-50
Km/ mg.mL^{-1}	4.078	0.746	3.688
Vmax/mg.mL^{-1}.min^{-1}	0.127	0.0268	0.065
Metal activator	Mn^{2+},Mg^{2+}	Mn^{2+},Mg^{2+}	Ni^{2+},Co^{2+},Mn^{2+}
Metal inhibitors	Cu^{2+},Hg^{2+},Ag^+,Fe^{3+}	Ag^+,Hg^{2+},Pb^{2+}, Cu^{2+}	Fe^{3+},Sn^{2+},Pb^{2+},Hg^{2+}

Besides, Effect of metal ions on enzyme activity showed that both of the chitosanalytic and cellulolytic activities of CCBE were significantly inhibited by Hg^{2+}, activated by Mn^{2+},Mg^{2+}; and not significantly affected by other metal ions. The results revealed that the metal ions were not necessary for both chitosanolytic and CMCase activities of CCBE, which is similar with those of chitosanases and cellulases [43,59].

3.1.2. Purification and Characterization of a Bifunctional Chitosanase with Chitinase Activity from Commercial Lipase

Several authors investigated the effects of lipases on chitosan and chitin degradation [44, 53, 61, 62], but lipases these authors used were in crude form and it has not been determined whether these lipases were able to split both carboxyl ester bonds in acyl glycerol and β-1,4-glycosidic linkages in chitosan. In a previous study, we found that a commercial lipase preparation from *Aspergillus oryzae* could hydrolyze chitosan effectively [48].

On the basis of the established effective purification system of column chromatography, A hydrolase with chitosanolytic and chitinolytic activities but without lipolytic activity (CNBE) was purified from a commercial lipase preparation by using a combination of DEAE-Sepharose CL-6B anion exchange chromatography, hydrophobic interaction chromatography on a Phenyl-Sepharose CL-4B column, and gel filtration on a Sephacryl S-200 column through changing elution conditions [72]. And through these steps, the obtained enzyme was purified 33.8-fold with a recovery of 12.5%. Results of HPLC and SDS-PAGE analysis showed that the hydrolase CNBE had been purified to homogeneity. Because no other components of the lipase showed chitosanolytic activity during purification process, chitosanolytic activity the lipase exhibited was caused by the purified enzyme CNBE. Molecular mass of the purified enzyme estimated by SDS-PAGE and non-reducing SDS-PAGE ,gel filtration was about 74 kDa and 130 kDa(Fig2e), respectively, indicating that the enzyme was composed of two identical subunits bound together with disulfide bonds (Table 1).

The purified CNBE toward chitosan showed the optimum action pH value and temperature were 4.6 and 60°C, respectively, and it was stable in pH range of 4.5-9.5 and at temperatures lower than 60°C(seen in Table 2). Metal ions such as Ni^{2+}, Co^{2+}, and Mn^{2+} had obvious activation effects on the enzymatic activity, while Fe^{3+}, Sn^{2+}, Pb^{2+}, and Hg^{2+} inactivated the enzyme, and Na^+, K^+, Mg^{2+}, Zn^{2+}, and Cd^{2+} had no obvious effect on the purified enzyme. The Michaelis constant (Km) and maximum velocity (Vmax) calculated according to Lineweaver–Burk plots were 3.588 mg.mL^{-1} and 0.065 mg.mL^{-1}.min^{-1}, respectively.

There are several reports that the chitinases from plant tissues also showed chitosanase activity, which probably resulted from bioevolution of chitin metabolism of the chitinase-producing tissues [89,90]. Osswald et al isolated seven acidic chitinases isoforms (EC 3.2.1.14) from 4-week-old nonembryogenic *Citrus sinensis* L. Osbeck cv 'Valencia' callus tissue using size exclusion, anion exchange, and chromatofocusing HPLC techniques. Eleven isoforms were isolated with M_w between 26,000 and 37,400. Eight of the isoforms were purified to homogeneity and all but one cross-reacted with a polyclonal antibody raised against a basic class I potato leaf chitinase. The isoelectric points (determined by chromatofocusing) were from pH 4.5 to 5.4. All hydrolases degraded chitin and four were capable of hydrolyzing solubilized shrimp shell chitosan suggesting they may be chitosanases (EC 3.2.1.99). Apparent chitosanase activity generally decreased with decreasing acetylation of the chitosan (i.e. from 20% to 0% acetylation). The chitinases and chitinases/chitosanases are predominantly endochitinases. Chitosanase activity was optimal at pH 5 while the pH optimum for chitinase activity ranged between pH 3.5 and 5.5. The chitinases and chitinases/chitosanases wer stable up to 60°C and showed their highest enzyme activity at that temperature. N-terminal sequences were obtained on three of the isoforms. One of the isoforms was identified as a class II chitinase and the other two as class III chitinases.The bifunctional chitinases/chitosanases were the emergency products of plant defensing system [89]. Pozo also reported that new chitosanase acidic isoforms have been shown in *Glomus mosseae*-colonized tomato roots and their induction, together with the previously described mycorrhiza-related chitinase isoform, has been further corroborated in plants colonized with another *Glomus* species (*G. intraradices*),as well as in tomato roots colonized *in vitro* by *Giaspora rosea*. The induction of these chitosanase isoforms appears as a specific response to the arbuscular mycorrhizal (AM) symbiosis, and does not correspond to unspecific defence mechanisms, since these isoforms were not induced by the pathogen *Phytophthora parasitica*[90].

However, there are only two reports on the bifunctional enzymes with chitosanase and chitinase activity from non-specific chitosanolytic enzymes:one is our research from commercial lipase[72], and the other is from commercial stem bromelain[52]. The latter was purified from commercial stem bromelain by Hung etc through sequential steps of SP-Sepharose ion-exchange adsorption, HiLoad Superdex 75 gel filtration, HiLoad Q Sepharose ion-exchange chromatography, and Superdex 75 HR gel filtration. The purified hydrolase was homogeneous, as examined by sodium dodecyl sulfate-polyacrylamide gel electrophoresis. The enzyme exhibited chitinase activity for hydrolysis of glycol chitin and 4-methylumbelliferyl β-D-N,N',N-triacetylchitotrioside [4-MU-β-(GlcNAc)3] and chitosanase activity for chitosan hydrolysis. For glycol chitin hydrolysis, the enzyme had an optimal pH of 4, an optimal temperature of 60 °C, and a Km of 0.2 mg/mL. For the 4-MU-β-(GlcNAc)3 hydrolysis, the enzyme had an optimal pH of 4 and an optimal temperature of 50 °C. For the chitosan hydrolysis, the enzyme had an optimal pH of 3, an optimal temperature of 50 °C, and a Km of 0.88 mg/mL. For hydrolysis of chitosans with various N-acetyl contents, the enzyme degraded 30-80% deacetylated chitosan most effectively. The enzyme split chitin or chitosan in an endo-manner. The molecular mass of the enzyme estimated by gel filtration was 31.4 kDa, and the isoelectric point estimated by isoelectric focusing electrophoresis was 5.9. Heavy metal ions of Hg^{2+} and Ag^+, p-hydroxymercuribenzoic acid, and N-bromosuccinimide significantly inhibited the enzyme activity.Since the stem bromelain is abtained from plant,this bifunctional chitosanase/chitinase may be also attribute to the plant denfensing system,while it is still confused about the reason why lipase-producing *A. oryzae* also secreted bifunctional chitosanase-chitinase(CNBE),which needs a further study.

3.2. ACTION MODE ANALYSIS OF BIFUNCTIONAL ENZYMES ON CHITOSAN

3.2.1. Substrate Specificity

Chitosanases from various sources show different substrate specificity and hydrolytic action patterns, which are also dependent on the degree of polymerization and of acetylation of chitosan.

The activities of the purified bifunctional CCBE and CNBE toward various substrates are listed in Table 3. Although all chitosans with different DD used in the experiment were observed to be degraded by the two enzymes, the enzymes demonstrated different levels of activity on different chitosans. As shown in Table 3, CCBE purified from commercial cellulase could catalyze the hydrolysis of chitosan substrates having a wide range of acetyl groups; it showed similar activity on chitosan which was more than 70%DD, among these, it showed the best activity on 84%DD chitosan,while had no activities on colloid chitin and pNPG.

Table 3. Substrate specificity of the purified CCBE and CNBE

Chitosan (DD/%)	CCBE	CNBE
Chitosan (64%)	57.5	78.2
Chitosan (75%)	93	91.7
Chitosan (84%)	100	100
Chitosan (90%)	91	77.1
Chitosan (96%)	90	75.9

However, for CNBE purified from lipase, this enzyme exhibited higher activity toward chitosans with DD of 75% and 84%, lower activity toward those with DD of 64%, while showed lowest activity toward chitosans with DD more than 90%, which indicated that the CNBE degraded chitin more effectively than chitosan. In addition, when chitosans with the same DD were used as the substrates, chitosans with lower molecular mass were more susceptible to be hydrolyzed by the enzyme than those with higher DP (degree of polymerization)(data not shown).

3.2.2. Chitosan and Chitooligosaccharide Hydrolysis Analysis by CCBE

HPLC and Thin layer chromatography (TLC) are the common tools for products analysis of chitosan hydrolysis. Hydrolysis products generated by the actions of the purified CCBE on chitooligosaccharides are shown in Fig3. After reaction for 1 and 6 h, chitobiose, chitotriose, chitotetraose and chitopentaose were hydrolyzed and released glucosamine and other shorter chitooligos accharides. The products of chitosan and standard chitooligosaccharides digested with CCBE were also analyzed by HPLC/MS and time course analysis, besides TLC,indicating that CCBE could cleave GlcN-GlcNAc, GlcNAc-GlcN as well as GlcN-GlcN bonds from the non-reducing end of chitosan chain[56], which showed some difference with the split modes of specific chitosanases, for it is reported popularly that chitosanases can be classified into three main classes according to their substrate specificity: class I split GlcNAc-GlcN and GlcN-GlcN bonds, class II split only the GlcN-GlcN bonds, and class III chitosanases split GlcN-GlcN and GlcN-GlcNAc bonds[59].Besides, considering its action pattern of COS, the CCBE was identified as an exo-β-D-glucosaminidase, which releases glucosamine successively from the non-reducing ends of COS and hydrolyze chitosan and COS completely to monomers, glucosamines.

Insofar as we are aware, exo-β-D-glucosaminidase (GlcNase)has been purified and characterized from *Nocardia orientalis* [91], *T. reesei*[57], *Amycolatopsis orientalis* [92], *Penicillium funiculosum* [93], *Thermococcus kodakaraensis* [94], and *Aspergillus sp.* [95-98] and they were all induced by chitosan. Their molecular weights were generally between 90-135kDa with a exception of that from *A. flavus*(45kDa) and several GlcNases with trans glycosylase activity had been reported[92,93,97,98] while GlcNases from *N. Orientalis* [91] and *T. Reesei* [57] did not exhibit this kind of activity. However, our CCBE with a Mw of 66kDa was purified from a commercial cellulase, which produced by *T.viride* with the induction of cellulose. Moreover, CCBE also showed CMCase activity.TLC analysis indicated that CCBE degraded CMC to cellobiose and cellotriose, but had no activity on cellobiose, chitin-oligomers and pNPG,which suggested that CCBE acted CMC as a cellobiohydrolase,but no chitinase and transglycosylase activity(seen in table 6).

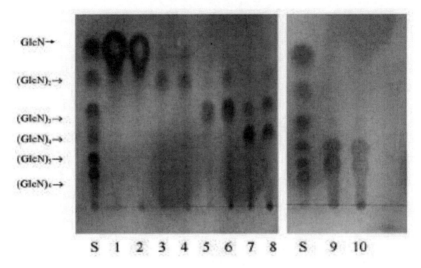

GlcN→
(GlcN)₂→
(GlcN)₃→
(GlcN)₄→
(GlcN)₅→
(GlcN)₆→

S 1 2 3 4 5 6 7 8 S 9 10

Figure 3. Thin layer chromatography of chitooligosaccharide hydrolysates by CCBE. Lanes 1, 3, 5, 7, 9 denoted glucosamine, chitobiose, chitotriose, chitotetraose and chitopentaose hydrolysis by the purified enzyme at 60 °C for 1 h. Lanes 2, 4, 6, 8,10 denoted glucosamine, chitobiose, chitotriose, chitotetraose and chitopentaose hydrolysis by the purified enzyme at 60 °C for 6 h. Lane S denoted standard chitooligosaccharide: glucosamine, chitobiose, chitotriose, chitotetraose, chitopentaose and chitohexose(Cited from reference[56,88]).

In our previous study, the products of chitosan hydrolysis by crude cellulase for 0.5–12 h were chitobiose, chitotriose, chitotetrose, and some chitooligasaccharides with long chain length. The shorter oligomer of D-glucosamine increased as digestion time increased (data nto shown), which indicated that the crude enzyme split chitosan in both exo-manner and endo-manner as indicated by the composition of chitosan hydrolysates. However, the purified enzyme, CCBE, one magical part of the crude enzyme, split chitooligosaccharides in an exo-manner and had high cellobiohydrolase activity as well, but no transglycosylation activity (data not shown). This property was consistent with the enzyme from *Myxobactor* AL-1 that exhibited both β- 1,4-glucanase and chitosanase activities. From the above, it can be seen that the bifunctional component with exo-β-D-glucosaminidase and cellobiohydrolase activity were the main mechanism of non-specific chitosan degradation of commercial cellulase from *T.viride* although there are still other endo-chitosanolytic components existing.Ike etc obtained a similar result from CBH I of *T.reesei* [99].

3.2.3. TLC Analysis of COS Hydrolysis Products by the Purified CNBE

The action mode of the purified CNBE was also studied by TLC and HPLC method through analyzing the hydrolysis products of standard compounds (chitosan-oligomers and chitin-oligomers) catalyzed by the purified enzyme [71,100]. The results turned out that the enzyme could not only hydrolyze chitosan oligomers with DP ranging from 2 to 6 (data not shown), but also degrade the chitin-oligomers.

The hydrolysis products of (GlcN)4 and (GlcN)5 by the CNBE were shown in Fig4A. At the early stage of hydrolysis of (GlcN)4, small quantities of GlcN and (GlcN)3 were produced; with the increase of the reaction time, (GlcN)2 was also generated; and in the end, the substrate (GlcN)4 and the intermediate products, i.e., (GlcN)3 and (GlcN)2, disappeared, while GlcN was the sole end product.For the hydrolysis of (GlcN)5, a similar results were observed (Fig. 4B). Firstly, the substrate was hydrolyzed to glucosamine and (GlcN)4, then (GlcN)4 was hydrolyzed to GlcN and (GlcN)3,and then (GlcN)3 was hydrolyzed to GlcN and (GlcN)2, and the final product was also glucosamine. These results suggested that this purified enzyme released glucosamine successively from the non-reducing ends of chitosan oligomers, and this is just the characteristic action of a GlcNase. In addition, no oligomers with DP higher than that of the substrates were produced during the course of hydrolysis (Fig4), indicating that this purified CNBE possessed no transglycosylase activity.

The hydrolysates of chitin-oligomers by CNBE were also analyzed by TLC (Fig4C and D).As seen in Fig4,the CNBE could degrade (GlcNAc)4 and(GlcNAc)6 to (GlcNAc)2,while had no reaction on (GlcNAc)2.This result suggested that CNBE could hydrolyse chitin as the N-acetyl-chitobiosidase pattern.

In conclusion, a chitosanolytic enzyme was purified to homogeneity from a commercial lipase preparation using a series of chromatographic separations and purifications, and this enzyme was then characterized as an exo-β-D-glucosaminidase which releases GlcN residues successively from (GlcN)n. In addition, it was also characterized as an N-acetyl-chitobiosidase which releases (GlcNAc)2 residues successively from (GlcNAc)n (seen in Table 4) .

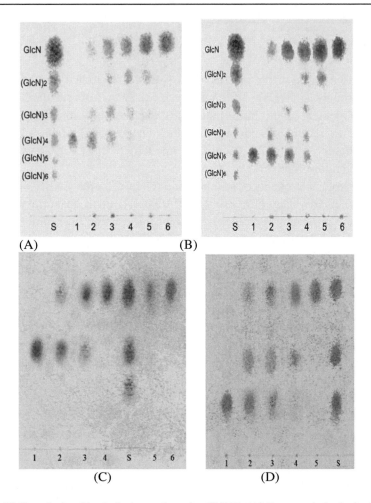

Figure 4. TLC analysis of hydrolysis products by CNBE. (A) Enzymatic hydrolysis of
(GlcN)4, Lane S denotes the standard chitosan oligomers. Lanes 1-6 denote products of
(GlcN)4 hydrolysis at 0 h, 0.5 h, 1 h, 3 h, 6 h, and 12 h, respectively. (B) Enzymatic
hydrolysis of (GlcN)5, Lane S denotes chitosan oligomers: glucosamine, chitobiose,
chitotriose, chitotetraose, chitopentaose, and chitohexaose. Lanes 1–6 denote products of
(GlcN)5 hydrolysis at 0 h, 0.5 h, 1 h, 3 h, 6 h, and 12 h, respectively.(C) hydrolysis
products of (GlcNAc)4 and (GlcNAc)2, Lanes 1–4 denote products of (GlcN Ac)4
hydrolysis at 0 h,1 h,6 h and12 h,respectively;Lane 5-6 denote products of (GlcN Ac)2
hydrolysis at 0 h and12 h.(D) hydrolysis products of (GlcNAc)6. Lanes 1–5 denote
products of (GlcN Ac)4 hydrolysis at 0 h,1 h,3h, 6 h and12 h,respectively. (Cited from
references[48,71,99]).

In the chapter, we purified a chitosanolytic enzyme free of lipase activity
from a commercial lipase preparation, which showed hydrolytic activity toward

chitosan in its crude form. This result shows that the chitosanolytic activity of this commercial enzyme was a result of the presence of the chitosanolytic enzyme, because this purified CNBE was the only chitosanolytic component of the crude enzyme and no other chitosanolytic activity was detected during the purification process.Therefore, the lipase, the essential component of the commercial lipase preparation,was not a bifunctional enzyme with both lipolytic activity and chitosanolytic activity. This result is similar to those of previous studies on several proteases [50-52].

Table 4. Action pattern analysis of the purified CCBE and CNBE

Substrates	CCBE	CNBE
CMC	Cellobiose,trimer,tetramer	--
Chitosan	Glucosamine	glucosamine
$(GlcN)_2$	glucosamine	glucosamine
$(GlcN)_{3-6}$	Glucosamine and $(GlcN)n-1$, $(GlcN)n-2$	Glucosamine and $(GlcN)n-1$, $(GlcN)n-2$
$(GlcNAc)_4$	--	$(GlcNAc)_2$
$(GlcNAc)_6$	--	$(GlcNAc)_2$
Glycosidic bonds	GlcN-GlcN、GlcNAc-GlcN 、 GlcN-GlcNAc	GlcN-GlcN、GlcN-GlcNAc、 GlcNAc-GlcN、GlcNAc-GlcNAc
Pattern	β-D-glucosaminidase /cellobiohydrolase	β-D-glucosaminidase /N-acetyl-chitobiosidase

For enzymes for commercial use, the purification process may be rather rough and proteins and enzymes with similar properties to the objective enzyme may be included in the preparation. Therefore, there is the possibility that the some chitosanolytic enzymes may be present in the crude lipase. According to the previous literature mentioned above and results of this present work, among the commercial enzymes which exhibit both their labeled activity and chitosanolytic activity, we speculate that carbohydrases such as pectinases and cellulases are more likely to be bifunctional enzymes than lipases and proteases, because the structures of the substrates of these carbohydrases are similar to each othe, which is confirmed from the experiments.

Chapter 4

STRUCTURE AND FUNCTION ANALYSIS OF BIFUNCTIONAL ENZYMES WITH CHITOSANOLYTIC ACTIVITY

4.1. AMINO ACID COMPOSITION OF PURIFIED CCBE AND CNBE

The Amino acid composition of two purified bifunctional enzymes CCBE and CNBE were analyzed by auto-amino acids analyzer. The results indicated that although the two enzymes had different amino acid composition, which announced they are different, both of them had higher concentrations of Asp and Glu, while those of His, Arg, and Lys were low, and those of sulfur-containing amino acids were extremely low.

4.2. STRUCTURE ANALYSIS OF PURIFIED CCBE AND CNBE

4.2.1. N-Terminal Sequencing Analysis of CNBE

The N-terminal sequence containing 12 amino acids of the purified CNBE from lipase were determined by Edman degradation sequencing,and N-terminal sequence of CNBE was tested as Ala – Leu – Arg – Leu – Asn – Ser – Pro – Asn – Asn – Ile – Ala – Val,which was blasted in the Non-redundant Protein Sequences database using the NCBI Blastp2 program. The results showed that this sequence was similar to that of a protein (XP_001824581) from A. oryzae, and the identity of the 12 amino acids was 100%. Since the protein XP_001824581 was identified as a chitinase belonging to glycoside hydrolases family 18(GH-18) according to the deduced amino acid sequence from its gene,it can be deduced that the purified CNBE prabably belonged to GH-18 and was identified as a chitinase with chitosanase activity. However, activity analysis indicated that the CNBE showed a little higher chitosanase activity than its chitinase activity, which means it is mainly as a chitosanase.According to the division of Glycoside hydrolases family [101,102], GH-18 is an ancient chitinase family,while the conserved chitosanases belongs to glycoside hydrolases family 46, 75 and 80, respectively. For recent years, most bifunctional enzymes with cellulolytic and chitosanolytic activity has been found belonging to glycoside hydrolase family 8 (GH-8), GH-5 and GH-7[60], which contains cellulase, xylanase, lichenase among others. Hence, the CNBE and GH-18 was a novel finding of the chitosanase family.

4.2.2. Sequencing Analysis of the Purified CCBE by MALDI-TOF Mass

Because of the block in the N-terminal of the purified CCBE, Matrix assisted laser desorption/ionisation time of flight mass spectrometry (MALDI-TOF mass) was used for sequencing of CCBE. The purified CCBE was separated by SDS-PAGE with Coomassie blue staining, and then in-gel digested by MALDI-TOF mass spectrometry. The resulting peptides of CCBE digests were analyzed on a Bruker AutoFlex MALDI-TOF-TOF mass spectrometer operating in MS and MS-MS mode in order to determine the peptide mass and peptide sequence.

Sequence analysis of purified CCBE using MALDI-TOF mass spectrometry was shown in Fig5 and Table5. The results indicated that the most significant match found was with cellobiohydrolase (CBH) from T.viride and T.reesei (gi|295937,gi|223874), demonstrating that the bifunctional fCCBE probably have cellobiohydrolase activity, which is consistent to the previous study[56].Since CBH gi|295937 and gi|223874 are part of glycoside hydrolase family 7(GH-7),we assumed that the CCBE belonged to GH-7. This was similar with the report of Ike et al[99]: they first found the CBH I of T.reesei had chitosanolytic activity and opened a novel chitosanase family—GH-7.Our finding of the CCBE from T.viride confirmed that GH-7 was a chitosanase family for the second time on earth. Accordingly, we can also infer that the bifunctional chitosanase-cellulases from fungi probably all belonged to GH-7, since those belonged to GH-8 and GH-5 all have been found from bacteria [60].

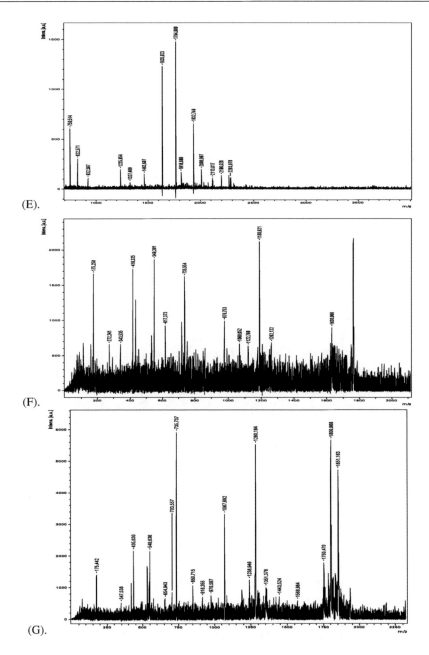

Figure 5. Maldi-TOF-TOF mass analysis of CCBE. (E), peptide mapping of CCBE digested by Trypsin; (F) and (G) are peptide mass fingerprint of the two main polypeptides 1764.0 and 1932.7488, respectively. (Cited from reference[142]).

Table 5. Sequencing of two main peptides of CCBE digested by trypsin

Observed	Mr(expt)	Mr(calc)	Peptide
1764.0005	1762.9932	1762.9577	K.KLTVVTQFETSGAINR.Y
1932.7488	1931.7415	1931.7203	R.YGGTCDPDGCDWNPYR.L

4.3. DETERMINATION OF THE ACTIVE SITES OF CCBE AND CNBE BY CHEMICAL MODIFICATION

From the above, the glycoside hydrolases belongings of the bifunctional CNBE and CCBE were primarily identified to GH-18 and GH-7, respectively, based on the N-terminal and Maldi-tof-tof mass sequencing. However, it is still not clear if the two bifunctional enzymes share one catalytic domain toward two substrates or not.

It is well-known that during the cellulose hydrolysis with cellulase, the active Glu and Asp residues are mainly involved in catalysis, while Trp,Tyr, His, Arg, Asn, etc., act as substrate-binding sites [88]. However, the essential active amino acids involved in specific chitosanases catalysis have seldom been reported, and the few known are similar to that of lysozyme, which has Glu and Asp in their catalytic core, although their locations in the enzymes differ. For example in a study on the active sites of chitosanases from Streptomyces sp. N174 and B.circulans MH-K1 using a combination of site-directed mutagenesis with thermal unfolding, fluorescence spectroscopy and X-ray crystallography, it was shown that Glu62 and Asp80, the highly conserved amino acids of N-terminal, were catalytic sites. Accordingly, Asp57 participated in substrate binding, and Asp205 combined with Asp145 and Asp190 to form a fixed framework which stabilized to the cleft of catalytic sites[103-106]. On the other hand, the catalytic sites of B. circulans MH-K1 chitosanase are Glu79 and Asp97, with Tyr148 and Lys218 involved in substrate binding[106,107].For chitosanase A from Matsuebacter chitosanotabidus 3001, Glu121 and Glu141 are catalytically important residues, and Asp139, Asp148, Arg150, Gly151, Asp164 and Gly280 are important for enzymatic activity, although they are not involved directly in catalysis[108].

The above information founded a basis for the determination of the active sites and relationships between structure and function of the bifunctional CCBE and CNBE and chemical modification of proteins has been a valuable technique to study structure and function of enzymes. Therefore, we have employed the specific chemical modifiers such as DEPC, NBS, Ch-T, EDAC, PMSF, CHD, NAI, DTNB, and DTT to determine the active sites of the CCBE and CNBE toward each substrate.

4.3.1. The Essential Carboxyl Group Modification using EDC

Participation of carboxyl groups in the catalytic function had been reported in a variety of glycosyl hydrolases [109-111]and the carboxylate seemed to be an integral part of catalytic site of these enzymes where acid/base catalysis was involved, since chemical labeling, site directed mutagenesis, X-ray crystallographic studies, and stereo chemical studies of the released hydrolytic products, all strongly suggested that during the cellulose hydrolysis by cellulase, the two carboxyl groups Glu, Asp were necessary in their catalysis; and the chitosanase, like hen egg white lysozyme, degraded chitosan involving two carboxyl residues in catalysis (one acting as the catalytic nucleophile and the other as the acid/base catalyst).These two carboxyl residues were totally conserved and were surrounded by a highly homologous region [112]. The CCBE, which was purified from a commercial cellulase prepared from *T. viride*, have been also reported to show different hydrolytic characteristics against the two substrates: Chitosan and CMC [56].However, it is still ambiguous whether the CCBE has two different active sites for its dual activities, for the active sites involved in conserved chitosanases and cellulases were not identical [60].

It is well-known that three essential carboxyl residues 212E, 214D and 217E are involved in the cellulose catalysis of fungi CBH I of GH-7, among these, 212E functions as a nucleophilic reagent, while 217E functions as a proton donor [113,114].Considering the CCBE was identified primarily to belong to GH-7 family, little is known about the enzyme-substrate interaction and the number of amino acids essential for catalytic activity of bifunctional chitosanase-cellulase enzyme (CCBE). In an attempt to assess if the catalytic mechanism of CBH I are functionally related to CCBE, and the role of carboxyl in CCBE catalysis, we have employed the water-soluble carbodiimide 1-ethyl-3-(3- dimethyl-aminopropyl) carbodiimide (EDC), a reagent which has bean widely used to modify carboxyl groups in enzymes for its esterifying reaction with the carboxyl side-chain of aspartate and glutamate residues as well as the C-terminal residue [115-119].

As a modifier, EDC is neither chromogenic by itself nor does it lead to such compounds; therefore, determination of the carboxylates with EDC is performed after kinetic analysis of the inactivation process by using different concentrations of modifier as follows [116]:

The inactivation reaction of the biocatalyst can be represented by

$$E + nI \xrightarrow{\ k1\ } EIn$$

Where E is the free enzyme, I the modifier, n the number of moles of modifier, and EIn the enzyme–modifier complex. The rate of the inactivation reaction is described by

$$U = -d[E]/d[t] = k1[E][I]_n$$

In most situations, when the modifier concentration greatly exceeds that of the biocatalyst, pseudo-first order kinetics is observed. For the pseudo-first-order rate, $k = k1 [I]n$, and after taking the logarithms one has

$$\log k = \log k1 + n\log[I]$$

If $T_{1/2}$ is the half-life of the enzyme, a plot of $\log(1/T_{1/2})$ versus $\log[I]$ should give a straight line with a slope equal to n, where n is equal to the number of modifier molecules (i.e., EDC) reacting with each active unit of the enzyme to produce an inactive EI complex [116].

According to the above, the nature of the catalytically-essential carboxyl groups in the CCBE and CNBE was delineated by chemical modification with EDC.

For CCBE, these experiments determined the role of carboxyl groups in the mechanism of CCBE acting on the chitosan and CMC, respectively. That the loss of the chitosanase and cellulase activity of CCBE caused by EDC was not coincided, suggested that carboxyl groups in CCBE played a dissimilar role on chitosan and CMC hydrolysis. According to the plot of the pseudo-first-order rate constants as a function of EDC concentration, we found that two carboxylates were essential for its chitosanase activity while one carboxyl group for its CMCase activity. Substrate protection experiments indicated that one of the two modified carboxyl groups reacted as the substrate binding site, and the other may be the catalytic group as the proton donor for CCBE's chitosanolytic activity. However, the CMCase activity of the enzyme was not protected against modification by CMC, indicating that the only modified carboxyl group was necessary for the catalytic activity of the CMCase, but not for the substrate binding.Since, EDC only activates unionized carboxyls [115], this means that the proton donating carxboxyl group in the active-site of both activities is modified at lower pH.

While for CNBE, when incubated with EDC for 10min, CNBE activity reduced rapidly, and lost 80% of its initial chitosanase activity with the 50 mmol/L of EDC. The higher the concentration of EDC was, the more evidently the activity decreased, which indicated that 1mol carboxyl group per CNBE existed in the catalytic domain of CNBE toward chitosan[100].

4.3.2. Modification of Tryptophan Residues using NBS

The indole side chain of tryptophan residues absorbs strongly near 280 nm. It is relatively reactive but is often protected from the solvent and reagents contained in the solvent by being buried in the interior of proteins along with other hydrophobic side chains. The lability of tryptophan to conditions normally used for acid hydrolysis of proteins has, in the past, made its quantitative determination relatively difficult. Treatment with N-bromosuccinimide is recommended to effect is selective modification in proteins. The indole group fo tryptophan residue could be inverted to hydroxide indole derivatives when reacting with NBS.the reaction is fast and unreversible, but could be tracked by the reduction of absorption in 280nm[120].the modified tryptophan residues were calculated according to the fomula:

$$ n = \frac{1.31 \times \Delta A_{280} \times Mw \times V}{m \times 5500} $$

Where n is the number of modified trp residues;1.31 is Witkop factor; ΔA_{280} is the reducing value of the absorption at 280nm;Mw is the molecular weight of the enzyme;V is the volume of the enzyme solution;m is the content of enzyme(g);5500 is the mol extinction coefficient of trp at 280nm.

Treatment with N-bromosuccinimide is usually done in acetate or formate buffers (pH 3 to 4), but values closer to neutrality can also frequently be used. Higher pH has an advantage in that little or no peptide bond cleavage takes place. Also higher pH is usually more selective, apparently differentiating between more and less reactive tryptophan residues. Reaction is usually relatively specific for tryptophan in the absence of sulfhydryl groups which are even more reactive.

In this present investigation, tryptophan residues on the CCBE and CNBE molecule were modified with NBS to assess the role of these residues in catalytic/substrate binding.The results of tryptophan modification by NBS showed that one Trp residue existed in the active site of CCBE's chitosanolytic activity,while two were involved in the active site of its CMCase: one located in the substrate binding domain,and the other located in the catalytic domain. Modification of enough Trp residues will lead to the completely loss in both activities of CCBE, suggesting that the tryptophan also help for the stability of enzyme conformation.

For CNBE, when the the moore retio of NBS and CNBE was 45.5, the activity was completely lost.However, chitosan showed significant protection on the CNBE, when preincubating with chitosan for 30min before NBS modification,the modified CNBE would retain 93% of its original activity,indicating that Trp was one of the essential catalytic amino acids and at least one mol Trp located in the substrate binding domain of CNBE.

4.3.3. Modification of Histidine Residue using DEPC

Diethylpyrocarbonate (DEPC) is a broad and highly sensitive histidine modifier. DEPC reacts with His residues in proteins to yield an N-carbethoxyhistidyl derivative, followed by an increase in absorbance at 240 nm [121]. The reaction is as follows:

Number of the modified His residues can be calculated from the absorption increase at 240 nm with a molar absorption coefficient of 3200M^{-1} cm^{-1} for N-carbethoxyhistidine. On the other hand, DEPC does not always react specifically with His but also with Tyr, Lys, Cys or carboxyl-containing residues in proteins [121]. Hydroxylamine has been shown to reverse the modified His and Tyr residues, which is much more reactive to the former than the latter, but does not reverse the modified Lys, Cys and carboxyl-containing residues.

The inactivation of the CCBE and CNBE by DEPC is ascribable to modification of His residues if the inactivation follows the absorption increase at

240 nm and exposure of the inactivated enzymes to hydroxylamine reverses the absorption at 240 nm accompanying regeneration of the activity.

The results indicated that each active site of the CCBE and CNBE had 1 histidine residue.For CCBE,the initial velocity of the two activities differ resulted from DEPC modification, potentially indicating that the two catalytic mechanisms of the dual activities were different.

4.3.4. Modification of Tyrosine, Threonine/Serine, Arginine, Methionine Residues Etc

N-Acetylimidazole (NAI) is a specific modifier of tyrosine residues. NAI acetylates the phenolic hydroxyl group in a reversible reaction, which reduced the absorption of enzyme at250-300 nm, especially at 278nm [122,123].Modification of tyrosine residues by NAI showed that tyrosine residue was essential for CCBE's chitosanase activity, but not involved in its CMCase catalysis.While the sharp decrease of both activities of CCBE by Ch-T were resulted from the crack of protein. In conclusion, the modified residues were all detected by activity and kinetic analysis, Ultra Voilet Spectroscopy, fluorescence, ative/SDS-PAGE and Circular Dichroism (CD) analysis.

Table 6. Comparison of amino acid composition of the two active centers of CCBE and CNBE

Residues	CCBE		CNBE
	chitosan （mol/mol CCBE）	CMC （mol/mol CCBE）	
Trp	1(substrate binding site)	2 (1mol was substrate binding site)	1(substrate binding site)
Cys	1	1	0
Ser/Thr	2	1	0
Tyr	1	0	0
His	1	1	1
Glu/Asp	2	1	1
Met	ND*	ND*	1

The results of chemical modification of CCBE and CNBE were seen in Table 6. As seen in Table 6, for CCBE, 1 mol Trp, 2 mol Thr/Ser, 1 mol Tyr, 1 mol His and 1 mol Glu/Asp per mol CCBE were the active residues involved in chitosanolytic activiy, while 1 mol Trp, 1 mol Thr/Ser, 1 mol Tyr, 1 mol His and

1 mol Glu/Asp per mol CCBE were essential for its cellulolytic activity. This result demonstrated that the purified CCBE had two distinct catalytic domains involved in chitosan and CMC hydrolysis, while part of the two catalytic domains probably overlapped [88,124].

For CNBE, modification of the purified CNBE by DEPC shows that 1 mole of histidine residue existed in the active site of the enzyme. Modification by NBS shows that tryptophan residues existed in the active site and there is at least 1 mole of tryptophan residue in the substrate binding site. Results of modification by Ch-T and EDAC indicate that methionine residues and carboxyl groups were essential groups of the purified enzyme. In addition, hydroxyl groups, arginine residues, tyrosine residues, sulfhydryl groups, and disulfide bonds were not essential groups of the purified enzyme [100].

Chapter 5

IDENTIFICATION OF CCBE FROM *T. VIRIDE* BY MOLECULAR CLONING

As we know, *T. viride* secretes at least two cellobiohydrolases (EC3.1.2.91), four endoglucanases (EC3.1.2.4) and two β-glucosidases (EC3.1.2.20). These enzymes have already been purified or their genes have been cloned [60], but none of them have been found showing chitosanase activity till now. Furthermore, Nogawa et al [57] and Duan et al [125] both found that *T. reesei* and *T.viride* could produce chitosanases with the induction of chitosan, respectively, but both of the two chitosanases had no CMCase activity. Hence, although we have found that the CCBE purified from the commercial cellulase preparation from *T.viride* exhibited both exo-β-D-glucosaminidase and cellobiohydrolase activity, it is necessary to use another method to prove its existence.

Nowadays, multidisciplinary approach combining proteomics with structural and functional genomics to identify novel proteins and define the structure and function of proteome have been a very useful and popular tool for proteomics research. To further identify if the bifunctional enzyme were produced by one single gene, we studied the gene structure and function of the CCBE choosing one *T.viride* strain by molecular cloning.

5.1. DETERMINATION OF FERMENTATION CONDITION OF T.VIRIDE

A cellulase hyper-producing *T.viride* strain was chosen for enzyme production. 10^6 conidia were inoculated into 100 ml Mandels medium [126], containing CMC (0.75%) rather than chitosan as the carbon source and were incubated for several days at 28 °C with vigorous shaking at 220rpm,and the temporal profile of chitosanase and cellulase was followed in liquid culture with CMC as the only carbon source during *T.viride* growth. The results showed that both of the two activities approached at its maximum on the 4th day. It is slightly dissimilar from the literature elsewhere which reported that the cellulase secreted by *T.viride* approached maximum on the 3rd day [127]. We once again came across to a conclusion that cellulase hyper-producing microorganisms could secrete chitoanolytic components in non-chitosan inducing conditions.

5.2. CLONING AND SEQUENCE ANALYSIS OF CCBE GENE

Taking the total RNA of T.viride as template,with the primers designed according to the Maldi-tof –tof mass result, the full-length cDNA sequence of CCBE was cloned by RT-PCR and SMART-RACE.Sequence analysis indicated that the full- length cDNA sequence of CCBE was identical to its DNA sequence, retaining two introns (67bp and 71bp, respectively). Blast search with the cloned cDNA sequence indicated that it had 88-100% identity with the four CBH I genes (Trichoderma. sp), while no similarity with chitosanase or chitinase genes reported in GenBank, which confirmed that the CCBE from T.viride belonged to GH-7,and its mRNA underwent alternative splicing .

In order to identify alternative splicing pattern of the CCBE gene,transcripts of CCBE from different developmental stages of *T.viride* during fermenting were analyzed for gene expressions by RT-PCR, as seen in Fig6. At the beginning of cultivation, there were no bands; after 2 -3days, three bands were observed and the two bigger bands were much more abundant than the smallest one. With increasing cultivation days, bigger bands gradually decreased to finally disappeared, while the contents of the smallest become much more abundant. Cloning and sequencing of these amplimers revealed that the three products at 1521bp, 1456/1460bp and 1389bp likely correspond to the CCBE splice variants ----DC, CCBEIN1/ CCBEIN2 (overlapped) and CC, respectively. The results indicated that the pre-mRNA splicing of CCBE gene in *T.viride* was a process, but not momentary, which is contrary to nature[128], and also underwent alternative splicing, suggesting the complexity in splicing and its regulation of fungi.

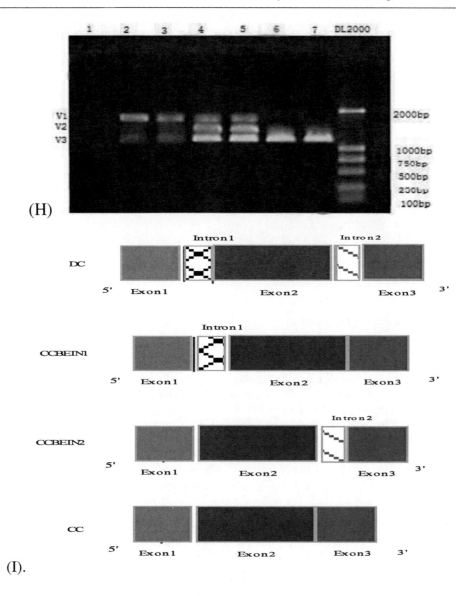

(H)

(I).

Figure 6. CCBE splice variants of in *T.viride* during cultivation using RT-PCR and their possible Gene structure from intron alternative splicing. (Cited from reference[142]). (H).PCR with primers located exon1 and 3 that flank the splice regions varied with the cultivating time.lane1-7 were cultivated for 1-7days.During cultivation, three bands at 1521bp, 1456/1460bp and 1389bp were observed that likely correspond to DC,CCBEIN1/CCBEIN2 (overlapped)and CC,respectively. DNA size markers were DL2000.(I).Gene structure of possible CCBE splicers.

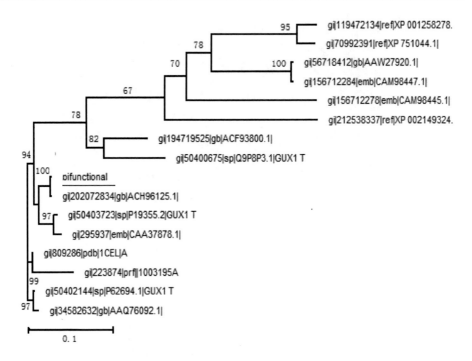

Figure 7. A phylogenetic tree constructed based on the deduced amino acid sequence of CCBE and those of fungal CBHs or exoglucanases of GH-7. Numbers at each branch indicate the percentage of times a node was supported in 1000 bootstrap pseudo replications by the Neighbor-Joining method. (Cited from reference[142]).

The assembled ORF of CCBE encodes a 514 amino acid polypeptide with a signal peptide of 17aa at the N-terminal. The mature protein was predicted to have a molecular mass of 54.2kDa and an isoelectric point (pI) of 4.4, showing insignificant difference with those of the purified CCBE(66kDa and 4.55)[56]. This was probably attributed to glycosylation modification of the fCCBE since it contained four N-glycosylation sites. The Sequence analysis by BLASTX program in NCBI database revealed that the putative amino acid sequence of CCBE shared high identities with known CBH I from *Trichodernma* sp., as *Trichoderma.sp* XST1 (ACH96125.1,99%), *T.viride* (P19355.2,91.1%; AAQ 76092, 91%; CAA37878,89.) *T.reesei* (1003195A,82.5%), *T.virens* (ACF93800, 80.9%), *T. harzianum* (Q9P8P3, 78.2%),while showed very lower identities with CBHs or exoglucanases from *Neosartorya sp.* (XP_001261835, 62.8%; XP_001258278, 62.6%), *Aspergillus sp*(XP_751044, 62%; XP_001214180, 59.5%), *Penicillium sp* (ACE60553, 61%; ACE60553, 60.4%), *Acremonium sp* (CAM98445, 61.2%),*Gibberella.sp* (AAS82858,56%). Alignment of the putative amino acid sequences from these sequences showed that CCBE contained two well-conserved motifs: a putative conserved domain of GH family 7 (position 19-440aa) and a fungal-fype cellulose-binding domain (fCBD)(position 481-514aa) conserved with four-cysteine cellulose-binding domain of fungi. According to the sequence-based glycosyl hydrolase classification [101,102]), we concluded that the CCBE belonged to family 7 of glycosyl hydrolases (GH-7).To analyze phylogenetic relationship between CCBE and CBHs and chitosanases from fungi, a Neighbor- Joining phylogenetic tree was constructed based on amino acid sequences of CCBE and other eight representative fungi (Figure 7), which showed that CCBE was the most related to *Trichoderma .sp* CBH(ACH96125.1) as they were clustered into the same clade, while it was distant to other CBHs from fungi species.

The above results indicated that CCBE belonged to GH-7,sharing high homology with CBH ofGH-7 family, but no similarity with conserved chitosanases from GH-46, GH-75,GH-80 and bifunctional chitosanases from GH-8[84,87] and GH-5[59,75,76]), which was in agreement with the results obtained from Maldi-tof Mass analysis of purified CCBE and reported by Ike et al[99], confirming that a novel chitosanase family—GH-7.However, an alternative splicing phenomenon existed in CCBE mRNA were probably the solo difference with that of CBH I. Hence, intron-retaining gene probably correlated with the dual functions of CCBE,the correlation was studied as follows through expression of different CCBE splicing genes in *pichia patoris* and properties analysis.

5.3. EXPRESSION ANALYSIS OF CCBE SPLICING GENES IN *PICHIA PATORIS*

Alternative splicing is the joining of different 5' and 3' splice sites, allowing individual genes to express multiple mRNAs that encode proteins with diverse and even antagonistic functions. Alternative splicing was first found in immunoglobulin μ gene[129] and Calcitonin mRNA[130] .Up to 59% of human genes generate multiple mRNAs by alternative splicing [131], and 80% of alternative splicing results in changes in the encoded protein[132], revealing what is likely to be the primary source of human proteomic diversity. Alternative splicing generates segments of mRNA variability that can insert or remove amino acids, shift the reading frame, or introduce a termination codon. Alternative splicing also affects gene expression by removing or inserting regulatory elements controlling translation, mRNA stability, or localization [133]. However, there are few reports exibiting chitosanases' and cellulases' genes existing alternative splicing phenomenon, only 5 cases.

Among these, 4 reports focused on alternative splicing of several cellulases. Sims etc first reported cellulase also underwent alternative splicing pattern.They found differential splicing of an intron in the CBH I.2 gene of Phanerochaete chrysosporium ME446; this intron lies within the region of the gene encoding the cellulose binding domain[134].Moreover, such differential splicing occurs in the CBH I gene of this fungus as well as in the CBH I gene and that this phenomenon is substrate dependent. Avicel elicits the synthesis of both classes of mRNA transcripts from both of these genes. In contrast, carboxymethyl cellulose predominantly elicits the synthesis of fully spliced transcripts from both genes. Such differential splicing might allow this fungus to regulate the specificities of substrate binding for these cellulases[135].In recent years,researchers found endoglucanases from *Mucor circinelloides*[136] and three oomycete plant pathogens *Phytophthora infestans,P. sojae and P. ramorum*[137], and CBH I from *Chrysosporium lucknowense*[138] also underwent alternative splicing phenomenon. Whereas, the ways of these alternative splicing varied with different sources. Baba etc cloned two cDNAs designated the mce1 and mce2 cDNAs from *Mucor circinelloides*, a member of the subdivision Zygomycota. The mce1 cDNA encoded an endoglucanase (family 45 glycoside hydrolase) having one carbohydrate-binding module (CBM), designated MCE1, and the mce2 cDNA encoded the same endoglucanase having two tandem repeated CBMs, designated MCE2. The two cDNAs contained the same sequences but with a 147-bp insertion. The corresponding genomic mce gene consisted of four exons. The mce1 cDNA was created from exons 1, 3, and 4, and the mce2 cDNA was created from exons 1, 2, 3, and 4. These results indicate that the mce1 and mce2 cDNAs were created from one genomic mce gene by "jumping" alternative splicing. MCE1 and MCE2, purified to apparent homogeneity from the culture supernatant of M. circinelloides, had molecular masses of 43 and 47 kDa, respectively. The carboxymethyl cellulase specific activity of MCE2 was almost the same as that of MCE1, whereas the Avicelase specific activity of MCE2 was two times higher than that of MCE1 when crystalline cellulose was abundant [136].Costanzo etc found many of the identified family 5 endo-(1–4)-β-glucanase genes (EGLs) copies were clustered in a few genomic regions of the oomycete plant pathogens *P. infestans*, *P. sojae*, and *P. ramorum*, and contained from zero to three introns. Using reverse transcription PCR to study in vitro and in planta gene expression levels of *P. sojae*, they detected partially processed RNA transcripts retaining one or more of their introns. In some cases, the positions of intron/exon splicing sites were also found to be variable. The relative proportions of these transcripts remain apparently unchanged under various growing conditions, but differ among orthogolous copies of the three Phytophthora species. The alternate processing of introns in this group of EGLs generates both coding and non-coding RNA isoforms. This is the first report on Phytophthora family 5 endoglucanases, and the first record for alternative intron processing of oomycete transcripts [137]. Furthermore, Gusakov etc also found a CBH I from Chrysosporium lucknowense, an industrial producer of cellulases and hemicellulases, underwent jumping alternative splicing. They purified two forms of cellobiohydrolase I (CBH I, Cel7A) from the culture ultrafiltrate of a mutant strain of the fungus C. lucknowense. The enzymes had different molecular masses (52 and 65 kDa, SDS-PAGE data) but the same pI (4.5). Peptide sequencing showed that a single gene encodes both proteins. Both enzymes displayed maximum activity at pH 5.0–5.5; they had similar specific activities against soluble substrates. However, the 65 kDa CBH I was much more efficient in hydrolysis of Avicel and cotton cellulose, and its adsorption ability on Avicel was notably higher in comparison to the 52 kDa enzymes. Using the in-gel tryptic digestion followed by MALDI-TOF mass-spectrometry, it was shown that the 52 kDa enzyme represents the core catalytic module of the intact 65 kDa CBH I without a cellulose-binding module and major part of glycosylated linker. The cbh1 gene was cloned and then the amino acid sequence of Cel7A was deduced from the gene sequence. The enzyme had high degree of similarity (up to 74%) to family 7 cellobiohydrolases and lower degree of similarity (up to 41%) to family 7 endoglucanases [138].Besides, it is reported that a HEX1 gene from a cellulase-hyper-producing fungi Trichoderma reesei strain exhibited intron-retaining alternative splicing[139].

With regard to chitosanases, Yamada reported a chitosanase gene(vChta-1) from Chlorella virus CVK2 had tissue-dependent alternative spicing.Western blot analysis with antisera raised against the vChta-1 protein identified two proteins of 37 and 65 kDa in virus-infected Chlorella cells beginning at 240 min postinfection and continuing until cell lysis. The larger protein was packaged in the virion, while the smaller one remained in the cell lysate. Both chitosanase proteins were produced from the single gene, vChta-1, by a mechanism of alternative gene expression [140], which is the first and the solo report about the alternative splicing of chitosanase genes.

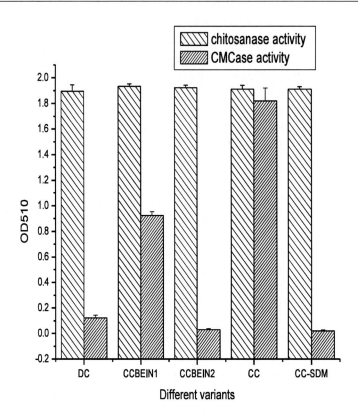

Figure 8. Activity analysis of CCBE different splicing variants expressed in *P.pastoris*. (Cited from reference[142]).

Up to now, **cellulases'** genes of *T.viride* have never been reported to undergo alternative splicing. This chapter found the CCBE gene from *T.viride* was actually a CBH I gene of GH-7 but had intron-retaining in its mRNA.Moreover, variant transcripts from the CCBE gene appear to be temporally alternatively spliced to generate mRNA that retained one or two introns at the middle stage of cultivation when the activity level was high, and the mature mRNA without introns were produced mainly at the end of cultivation when the quantity was high but activity decreased.Hence, to determine whether the retaining introns could influence dual function of CCBE, three splicing variants were assembled artificially in vitro [141]and expressed in *P.pastoris*. Activity analysis showed that all of the splicers had comparable strong chitosanase activity, while their cellulase activity varied (seen in Fig 8). CCBEIN1 had moderate CMCase activity compared to its chitosanase activity, and the dual activities of CC were comparative, however, CCBEIN2 and DC almost had no CMCase activity.

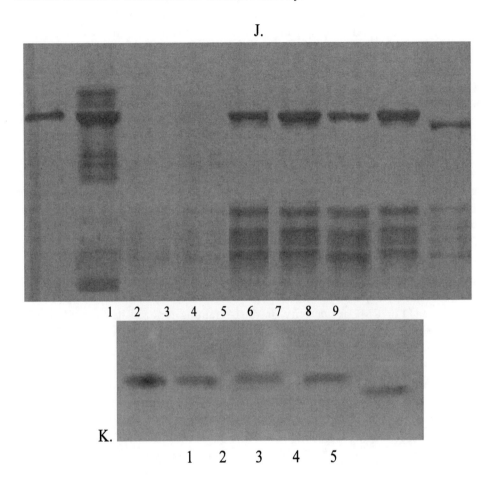

Figure 9. SDS-PAGE(J) and Western blotting (K) analysis of wild and different recombinant enzymes using 10%gel.(J).Lane1.purified CCBE; Lane 2. Fermenting solution of T.viride; Lane3-9 represents the blank control,Ppic9k, DC,CCBEIN1, CCBEIN2, CC and CC-SDM1expressed in *P.pastoris*; (K). Lane 1-5 were Western blotting analysis of DC, CCBEIN1, CCBEIN2, CC and CC-SDM1 expressed in *P.pastoris*,respectively. (Cited from reference[142]).

SDS-PAGE analysis of the recombinant splicing variants (Fig 9) revealed that all the bands of recombinant variants were in the same line with the purified CCBE, the molecular weights of them were 66kDa. Besides, total RNA was also isolated from the four recombinant *P.pastoris* strains (DC,CCBEIN1,CCBNIN2 and CC) and used as a template for RT-PCR,the results revealed that all the RT-PCR products were the same size and their sequences were identical without introns retaining(not shown). From which, we can deduce that all the intron-retaining variants could be spliced correctly to remove the introns in *P.pastoris* and produced proteins with the same amino acid sequence ,while their protein foldings were probably changed due to the intron-retaining influence. Hence, we convinced that our CCBE belonged to GH-7; the dual functions of CCBE were originated one gene, but their catalytic domains differ; and its gene underwent alternative splicing phenomenon, the retained introns had no significant effects on its chitosanase activity, but affected its cellulolytic function directly. Moreover, the relationship between the gene structure and bifunctional properties of CCBE were also discussed through the characterizations, kinetics and action modes analysis. The results showed that all the recombinant variants were similar in optimal hydrolysis conditions: between 55-60°C, and pH 5.0-5.2 toward chitosan; while toward CMC, the CCBEIN1 and CC showed the same hydrolyzing characteristics with that of purified CCBE. While, the retained introns made their proteins enhance the optimum pH range for two substrates hydrolysis as well as the thermal-stability of its chitosanase activity. The retaining of intron 1 increased the chitosanase activity more significantly, while decreased the affinity of CCBE to chitosan. For CMCase activity, intron1-retaining reduced its thermal-stability but had no effects on the cellulose-binding ability, which indicated that CCBE had two different binding domains for the two substrates.

Product analysis indicated that the action modesof four recombinant enzymes toward chitosan were identical to cleavage GlcN in the non-reducing terminal (data not published),however, the optimum substrate specifity of the recombinant enzymes were varied with different introns retaining. The recombinant CCBEIN1 and CC predominantly degraded COS with higher DP, while CCBEIN2 and DC preferentially degraded the smaller one. The C-terminal deletion mutation confirmed that the conformation of CBD changed with the intron-2-retaining to make the CMCase activity lost. From the above, it could be seen that the catalytic and substrate-binding domains of CCBE on chitosan hydrolysis located in the in the upper-middle reaches of N-terminal, which is different with that of CMC. Besides,a C-terminal deletion mutant,CC-SDM1, showed no change in its chitosanase activity, but lost its more than 80% CMCase activity(Fig8), and its molecular weight became to 55kDa(Fig9), suggesting that the cellulose binding domain was essential for the CCBE's cellulolytic activity, but had no effects on its chitosanase activity[142].

From the above results, it is confirmed that the dual activities of CCBE were determined by one single gene and had two distinct catalytic domains.This CCBE gene from *T.viride* underwent an intron-retaining alternative splicing pattern, which is dependent on the inducing periods. Although there were four variant splicing transcripts, the expressed proteins shared one amino acid sequence. The retaining introns affected the folding of proteins by changing their conformations to vary their functions. Intron1 changed the folding of catalytic domains slightly, while intron2 made its cellulase activity lost probably due to change the folding conformation of cellulose-binding domain to block the entrance of cellulose, and further interfere the binding of cellulose, which is confirmed by the C-terminal deletion analysis. However, the catalytic domain of CCBE on chitosan was probably located near the N-terminal or the upper part of the sequence which was influenced slightly by retaining introns. Further research was being done to make sense of the catalytic domain of GH-7 chitosanase on chitosan and the chitosanase-producing mechanism of hyper-cellulolytic fungus since the structural and functional relationship of CCBE of GH-7 is still not clear.

ACKNOWLEDGMENTS

This chapter was financially supported by National Nature Science Foundation of China (NSFC, No.20271023, No.20576104, and No.20876068), SKLF-MB-200805 and the State High-tech project (863) (No.:2007AA100401 and No.:2006AA09Z444).We also gratefully acknowledge the other contributors: Dr.Su C., Dr Liu J., and Dr. Lee D.X.

REFERENCE

[1]   Shahidi, F., Arachchi, J., & Jeon, Y. J. (1999). Food applications of chitin and chitosans. *Trends Food Sci Technol. 10*, 37–51.

[2]   Hong, S.P., &Kim, D.S. (1998).Chitosanolytic characteristics of cellulases from *Trichoderma viride* and *Trichoderma reesei. Korean J Food Sci. Technol.30,*245-252.

[3]   Hu, Y., Du, Y.M., &Yang,J.H., et al.(2007).Self-aggregation and antibacterial activity of N-acylated chitosan *.Polymer.8,*3098-3106.

[4]   Xu, J.G., Zhao, X.M., &Han, X.W. (2007).Antifungal activity of oligo chitosan against *Phytophthora capsici* and other plant pathogenic fungi in vitro. *Pesti Biochem Physiol.87,* 220-228.

[5]   Tsai,G.J., Zhang S.L., & Shieh P.L.(2004). Antimicrobial activity of a low-molecular –weight chitosan obtained from cellulase digestion of chitosan. *J Food Protec. 67,*396-398a.

[6]   Chung, Y.C., &Chen, C.Y.(2008). Antibacterial characteristics and activity of acid-soluble chitosan. *Bioresource Technol. 99,* 2806–2814.

[7]   Wei, X.L., &Xia, W.S.(2003). Research development of chitooligo saccharides Physiological activities.*China Pharma. Bulletin. 19(6),*614-617.

[8]   [8]Lin, S.B., Lin, Y.C., & Chen, H.H.(2009). Low molecular weight chitosan prepared with the aid of cellulase, lysozyme and chitinase: Characterisation and antibacterial activity. Food Chem.116 (1), 47-53.

[9]   Qin, C.Q., & Du, Y.M.(2002). Enzymic preparation of water-soluble chitosan and their antitumor activity. *Internet. J. Biol. Macromol.31,* 111–117.

[10] Qin, C.Q., Zhou, B., & Zeng, L.T., et al.(2004). The physicochemical properties and antitumor activity of cellulase-treated chitosan. *Food Chem. 84,* 107–115.

[11] Huang, R.H., Mendis, E., & Rajapakse, N. ,et al.(2006).Strong electronic charge as an important factor for anticancer activity of chitooligo saccharides (COS).*Life Science.78(20),* 2399-2408.

[12] Wang, S.L., Lin, H.T., & Liang, T.W.,et al.(2008). Reclamation of chitinous materials by bromelain for the preparation of antitumor and antifungal materials. *Bioresource Technol. 99(10),* 4386-4393.

[13] Ngo, D.N., Kim, M.M., &Kim, S.K. (2008).Chitin oligosaccharides inhibit oxidative stress in live cells. *Carbohydr Polym.74(2),*228-234.

[14] Ngo,D.N., Lee,S.H., & Kim, M.M., et al.(2009). Production of chitin oligosaccharides with different molecular weights and their antioxidant effect in RAW 264.7 cells *.J Functional Foods.1(2),*188-198.

[15] Okamoto, Y., Inoue ,A., & Miyatake, K. et al.(2003). Effects of chitin/chitosan and their oligomers/monomers on migrations of macrophages. *Macromolecular Biosci. 3,*587–590.

[16] Zaharoff, D.A., Rogers, C.J., & Hance, K.W., et al.(2007). Chitosan solution enhances both humoral and cell-mediated immune responses to subcutaneous vaccination. *Vaccine. 25,* 2085–2094.

[17] Feng, J., Zhao, L., & Yu, Q. (2004).Receptor-mediated stimulatory effect of oligochitosan in macrophages.*Biochem Biophys Res Commun.317,* 414–420.

[18] Kim, S. K., & Rajapakse, N. (2005). Enzymatic production and biological activities of chitosan oligosaccharides (COS): A review. *Carbohydr Res.62,* 357–368.

[19] Dou, J.L., Tan, C.Y., &Du, Y.G.,et al.(2007). Effects of chitooligos accharides on rabbit neutrophils in vitro.*Carbohydr. Polym. 69,* 209–213.

[20] Je, J.Y., Kim, E.K., &Ahn, C.B., et al.(2007). Sulfated chitooligo saccharides as prolylendopeptidase inhibitor. *Inter. J. Biol. Macromol. 41,* 529–533.

[21] Park, P.J., Je, J.Y., &Kim,S.K.(2003).Angiotensin I Converting Enzyme (ACE) Inhibitory Activity of Hetero-Chitooligosaccharides Prepared from Partially Different Deacetylated Chitosans. *J. Agric. Food Chem. 51 (17),* 4930–4934

[22] Ngo, D.N., Qian, Z.J., & Je, J.Y. , et al.(2008). Aminoethyl chitooligosaccharides inhibit the activity of angiotensin converting enzyme. *Process Biochem.43(1),* 119-123.

[23] Park, J. K., Shimono, K., &Ochiai, N., et al. (1999). Purification, characterization, and gene analysis of a chitosanase (ChoA) from the *Matsuebacter Chitosanotabidus* 3001. *J Bacteriol. 181,* 6642–6649.

[24] Abdel-Aziz,S.M., Mostafa,Y.A., & Moafi,F.E.(2008).Partial Purification and Some Properties of the Chitosanases Produced by *Bacillus Alvei* Nrc-14. *J Applied Sciences Res. 4(10),*1285-1290.

[25] Chen, X., Xia, W., & Yu, X. (2005). Purification and characterization of two types of chitosanase from *Aspergillus* sp. CJ22-326. *Food Res Inter. 38,* 315–322.

[26] Zhang, X. Y., Dai, A. L., &Zhang, X. K., et al. (2000). Purification and characterization of chitosanase and exo-β –D glucosaminidase from a Koji Mold, *Aspergillus oryzae* IAM2660. *Biosci Biotechnol Biochem.64,* 1896–1902.

[27]]Ouakfaoui, S. E., & Asselin, A. (1992). Diversity of chitosanase activity in cucumber.*Plant Science. 85,* 33-41.

[28] Lin, Q., & Ma, K.L.(2003). Study of catalytic hydrolysis of chitosan by cellulase. *China surfactant deter Cosmetics.33*, 22-25.

[29] Liu, J., & Xia, W.S.(2005). Characterization of chitosanase from cellulase produced by *Trichoderma viride.Chinese J Biochem Mol Biol.21,*713-716.

[30] Liu, Y.J., Jiang, Y., &Feng, Y.F., et al.(2005). Study on the chitosan hydrolysis catalyzed by special cellulase and preparation of chitooligo saccharide. *J func polym.18,* 325-329.

[31] Chen, J.Y., &Wu, G.M.(2003). Effect of cellulase on chitosan hydrolysis .*J. Guangdong. Medical College. 21*, 105-107.

[32] Kittura, F.S., Acharya, B., &Kumara, A.B.V., et al.(2003).Low molecular weight chitosan preparation by depolymerization with *Aspergillus niger* pectinase, and characterization. *Carbohydr Res.338,* 1283-1290.

[33] Kittura, F.S., Kumara, A.B.V., &Gowdab, L.R.,et al.(2003). Chitosanolysis by a pectinase isozyme of *Aspergillus niger*-A non-specific activity. *Carbohydr Polym.53,* 191–196.

[34] Sardar, M., Roy, I., & Gupta, M.N.(2003).A smart bioconjugate of alginate and pectinase with unusual biological activity toward chitosan. *Biotech Prog. 19*, 1654-1658.

[35] Shinya, Y., Lee, M.Y., &Hinode, H.,et al(2001).Effects of N-acetylation degree on N-acetylated chitosan hydrolysis with commercially available and modified pectinases. *Biochem Eng J.7*, 85–88.

[36] Roy, I., Sardar, M., &Gupta, M.N. (2003). Hydrolysis of chitin by pectinexTM.*Enzyme Microbial Technol.32*, 582–588.

[37] Roncal,T., Oviedo, A., &Armentia, I.L.D.,et al.(2007). High yield production of monomer-free chitosan oligosaccharides by pepsin catalyzed hydrolysis of a high deacetylation degree chitosan. *Carbohydr Res.342,* 2750-2756.

[38] Su, C., Xia, W.S., &Yao, H.Y.(2002). The relationship between structure of chitosan and papain activity. *J. Wuxi Univer. light Indus.21*, 112-115.

[39] Muzzarelli, R., Terbojevich, M., &Muzzarelli, C., et al.(2002). Chitosans depolymerized with the aid of papain and stabilized as glycolsylamines. *Carbohydr Polym.50,* 69–78

[40] Huang, Y.C., Li, L., & Guo, S.Y.,et al.(2003). Characteristics of chitosan degrading by papain. *J South China Univer Technol (nature science edition).31*, 71-75.

[41] Muzzarellia, R., &Tomasetti, M.(1994).Depolymerization of chitosan with the aid of papain. *Enzyme Microb Technol.16(2),* 110-114.

[42] Terbojevich, M., Cosani, A., &Muzzarelli, R.A.A.(1996).Molecular parameters of chitosans depolymerized with the aid of papain.*Carbohydr Polym. 29(1)*, 63-68.

[43] Zhou, G., He, Z.P., & Deng, G.H., et al.(2003). Enzyme kinetics of amylase and cellulase on hydrolyzing chitosan. *China Marine Sci. 27*, 59–63.

[44] Muzzarelli, R., Xia, W., &Tomasetti, M., et al.(1995). Depolymerization of chitosan and substituted chitosans with the aid of a wheat germ lipase preparation. *Enzyme Microbial Technol. 17*, 541–545.

[45] ia, W. S., & Muzzarelli, R. (1996).Study on the degradation of chitosan and its derivatives by lipase.*J.Wuxi. Univers. Light. Industry. 15(1)*,325-332.

[46] hou, S.Y., Yu, P., &Chen, S.,et al. (2003). Studies on the catalytic degradation of chitosan by lipase. *J. Fujian. Medical Univer.* 36, 302–305.

[47] Wang, L.J., Xia, W. S., &Chen,X.E. (2004).Study on the chitosan hydrolysis by lipase.*Food Industrial Sci.Technol. 25(5)*,302-304.

[48] Lee, D.X., Xia, W. S., & Zhang, J.L. (2008). Enzymatic preparation of chitooligosaccharides by commercial lipase. *Food Chem, 111*, 291–295.

[49] Kumar, V. A.B., Gowda, L.R., &Tharanathan, R.N.(2004). A comparative study on depolymerization of chitosan by proteolytic enzymes. *Carbohydr Polym.58*,275-283.

[50] Chiang, C. L., Chang, Y. M., & Chang, C. T., et al.(2005). Characterization of a chitosanase isolated from a commercial ficin preparation. *J Agri Food Chem. 53*, 7579–7585.

[51] Fu, J. Y., Wu, S. M., & Chang, C. T., et al.(2003). Characterization of three chitosanase isozymes isolated from a commercial crude porcine pepsin preparation. *J Agri Food Chem. 51*, 1042–1048.

[52] Hung, T. H., Fu, J. Y., & Chiang, C. L., et al.(2002). Purification and characterization of hydrolase with chitinase and chitosanase activity from commercial stem bromelain. *J Agri Food Chem.50*, 4666–4673.

[53] Pantaleone, D., Yalpani, M., & Scollar, M. (1992). Unusual susceptibility of chitosan to enzymatic hydrolysis. *Carbohydr Res. 237*, 325–332.

[54] Yalpani, M., & Pantaleone, M. (1994). An examination of the unusual susceptibilities of aminoglycans to enzymatic hydrolysis. *Carbohydr Res. 256*, 159–175.

[55] Somashekar, D., & Joseph, R.(1996). Chitosanase–properties and applications:a review. *Bioresource. Technol. 55*, 35–45.

[56] Liu, J., &Xia, W.S.(2006). Purification and characterization of a bifunctional enzyme with Chitosanase and cellulase activity from commercial cellulase. *Biochem. Eng. J. 30*, 82-87.

[57] Nogawa, M., Takahashi, H., & Kashiwagi, A., et al.(1998).Purification and characterization of exo-β-D-Glucosaminidase from a cellulolytic fungus, *Trichoderma reesei* PC-3-7.*Appl. Environ. Microbial.64*, 890-895.

[58] Hedges, A., &Wolfe, R.S.(1974).Extracellular enzyme from Myxcobacter AL-1 that exhibits both β-1,4-glucanase and chitosanase activities. *J.Bacteriol.120(2)*, 844–853.

[59] Kim, P., Tae, K.H., &Chung, K.J., et al.(2004). Purification of a constitutive chitosanase produced by Bacillus sp. MET 1299 with cloning and expression of the gene. *FEMS Microbiol. Lett. 240*, 31–39.

[60] Xia, W.S.,Liu,P., &Liu,J. (2008).Advance in chitosan hydrolysis by nonspecific cellulases. *Bioresource technol. 99(15)*,6751-6762.

[61] Sashiwa, H., Fujishima, S., &Yamano, N., et al.(2003). Enzymatic production of N-acetyl-D-glucosamine from chitin.Degradation study of N-acetyl chitooligosaccharide and the effect of mixing of crude enzymes. *CarbohydrPolym.51*, 391–395.

[62] Shin, S. S., Lee, Y. C., & Lee, C. (2001). The degradation of chitosan with the aid of lipase from Rhizopus japonicus for the production of soluble chitosan. *J Food Biochem. 25*, 307–321.

[63] Woolley, P., &Petersen, S.B. (1994). *Lipases: their structure, biochemistry, and application.* Cambridge: Cambridge University Press.

[64] Ma,R., &Huang,M.Z. (2002).Study on the Kinetics of the Lipase- promoted Hydrolysis of Chitosan.*Chemical World.43(9)*,472-475.

[65] Anthonsen, M. W., Varum, K. M., & Smidsrod, O. (1993). Solution properties of chitosans: Conformation and chain stiffness of chitosans with different degrees of N-acetylation. *CarbohydrPolym. 22*, 193–201.

[66] Anderson, J. W., Nicolosi, R. J., & Borzelleca, J. F. (2005). Glucosamine effects in humans: A review of effects on glucose metabolism, side effects, safety considerations and efficacy. *Food Chem Toxicology. 43*, 187–201.

[67] Zhou,G.,Tan,X.C., &Huang,Z.Y.,et al. (2002). Hydrolysis of chitosan by trypsin. *J.Guangxi Agri Biol Sci. 21(1)*,50-53.

[68] Hung, T. H., Fu, J. Y., &Chiang C L, et al.(2002). Purification and characterization of hydrolase with chitinase and chitosanase activity from commercial stem bromelain . *J Agri Food Chem.50*, 4666-4673.

[69] Muzzarelli, R. A. A., Tomasetti,M., &Ilari,P. (1994).Deploymerization of chitosan with the aid of papain. *Enzyme Microbial Technol.16(2)*,110-114 .

[70] Lin, H., Wang, H.Y., &XueC.H., et al.(2002). Preparation of Chitosan Oligomers by Immobilized Papain. *Enzyme Microbial Technol. 31 (5)*, 588-592.

[71] Xia, W.S., &Lee, D.X.(2008).Purification and characterization of exo-β-d-glucosaminidase from commercial lipase. *Carbohydr Polym.74(3)*,544-551.

[72] Tan,J.,Chen J.W., &Xia,W.S.,et al.(2007).Separation of papain with the chitosanase activity by ultrafiltration.*Food machinery.23(6)*,20-23.

[73] Macarron, R., Acebal, C., &Castillon, M. P., et al.(1996).Mannanase activity of endoglucanase III from *Trichoderma reesei* QM9414. *Biotechnol Letters.18*, 559–602.

[74] Hashimoto, W., Miki, H., &Wanakai, H., et al.(1998). Molecular cloning of two genes from β-D-glucosidase in *Bacillus sp.* GL1 and identification of one as a gellan-degrading enzyme. *Archieves Biochem Biophysics. 360*, 1–9.

[75] Pedraza-Reyes, M.(1997).The bifunctional enzyme chitosanase-cellulase produced by the gram negative microorganism *Myxobater.sp.*AL-1 is highly similar to Bacillus subtilis endo- glucanases. *Arch Microbiol.68*, 321-327.

[76] Tanabe, T., Morinaga, K., &Fukamizo, T.(2003).Novel chitosanase from *Streptomyces griseus* HUT 6037 with transglycosylation activity. *Biosci Biotech Biochem. 67*,54-364.

[77] Ohtakara, A., & Ogat, H. (1984). Purification and characterisation of chitosanase from *Streptomyces griseus*. Ed. Zikakis ,J. P. In *Chitin, Chitosan and Related Enzymes*, (147-159), Academic Press.

[78] Pelletier, A., & Sygusch, J. (1990).Purification & characterization of three chitosanase activities from *Bacillus megaterium* Pl. *Appl Environ Microbial. 56(4)*, 844-848.

[79] Mitsutomi,M.,Isono, M., & Uchiyama, A.,et al.98).itosanase activity of the enzyme previous reported as β-1,3/ β-1,4-glucanase from *Bacillus circulans* WL-12. *Biosci Biotech iochem, 62(11)*, 2107-2114.

[80] Uchida, Y., Izume, M., & Ohtakara, A. (1989). Purification and enzymatic properties of chitosanase from *Bacillus sp.* No. 7-M. *Bulletin Faculty Agri.66*,105–116.

[81] Kurakake, M., Yo-u, S., & Nakagawa, K.,et al. (2000).Properties of Chitosanase from B*acillus cereus* S1. *Current Microbiol.40*,6–9.

[82] Choi, Y.J., Kim, E.J., & Piao, Z.Y.,et al. (2004).Purification and characterization of chitosanase from *Bacillus sp.*strain KCTC 0377BP and its application for the production of chitosan oligosaccharides. *Appl Environ Microbe. 70*,4522- 4531.

[83] Hong, I.P., Jang, H.K., & Lee, S.Y, et al. (2003).Clone and characterization of a bifunctional cellulase-chitosanase gene from *Bacillus licheniformis* NBL420.*J Microbiol Biotechnol. 13*,35-42.

[84] Gao, X.A., Ju, W.T., & Jung,W.J.,et al.(2008). Purification and character ization of chitosanase from *Bacillus cereus* D-11. *Carbohydr Polym.72 (3)*,513-520.

[85] Jang, H.K., Yi, J.H., & Kim, J.T., et al.(2003).Purification, characterization, and gene cloning of chitosanase from *Bacillus cereus* H-1. *J Korean Microb Biotech. 31*, 216-223.

[86] Su, C.X., Wang, D.M., & Yao, L.M., et al.(2006).Purification, characterization, and gene cloning of a chitosanase from *Bacillus species* strain S65.*J Agri Food Chem. 54*, 4208-4214.

[87]]Ogura, J., Toyoda, A., & Kurosawa, T ., et al. (2006).Purification, characterization and gene analysis of cellulase (Cel8A) from *Lysobacter sp*.IB-9374. *Biosci Biotech Biochem.70,*2420-2428.

[88] Liu, J.(2006).The mechanism of chitosan hydrolysis by cellulase.*phD Dissertation*.Jiangnan University.Wuxi.

[89] Osswald, W. F., Shapiro, J. P., & Doostdar, H., et al. (1994).Identification and characterization of acidic hydrolases with chitinase and chitosanase activities from sweet orange callus tissue . *Plant Cell Physiol.35*,811-820.

[90] Pozo, M., Azcon-Aguila, C., & Dumas-Gaudot, E.,et al.(1998).Chitosanase and chitinase activities in tomato roots during interactions with arbuscular mycorrhizal fungi or *Phytophthora parasitica.J Experimen Botany.49*, 1729-1739.

[91] Nanjo, F., Katsumi, R., & Sakai, K. (1990). Purification and characterization of an exo-β-D-glucosaminidase, a novel type of enzyme, from *Nocardia orientalis. J Biol Chem.265*, 10088–10094.

[92] Cote, N., Fleury, A., & Dumont-Blanchette, E., et al.(2006). Two exo-β-D-glucosaminidase/ exochitosanases from actinomycetes define a new subfamily within family 2 of glycoside hydrolases. *BiochemJ.394*, 675–686.

[93] Matsumura, S., Yao, E., & Toshima, K. (1999). One-step preparation of alkyl β-D-glucosaminide by the transglycosylation of chitosan and alcohol using purified exo-β-D-glucosaminidase. *Biotechnol Letters. 21*, 451–456.

[94] Tanaka, T., Fukui, T., & Fujiwara, S., et al(2004). Concerted action of diacetylchitobiose deacetylase and exo-β-D-glucosaminidase in a novel chitinolytic pathway in the hyperthermophilic archaeon *Thermococcus kodakaraensis* KOD1. *J Biol Chem.279*, 30021– 30027.

[95] Zhang, X. Y., Dai, A. L., & Zhang, X. K., et al.(2000). Purification and characterization of chitosanase and exo-β-Dglucosaminidase from a Koji Mold, Aspergillus oryzae IAM2660. *BioscieBiotechnol Biochem. 64*, 1896–1902.

[96] Ji, J. H., Yang, J. S., & Hur, J. W. (2003). Purification and characterization of the exo-β-D-glucosaminidase from *Aspergillus flavus* IAM2044. *J Microbiol Biotechnol. 13*, 269–275.

[97] Jung, W. J., Kuk, J. H., & Kim, K. Y.,et al. (2006). Purification and characterization of exo-β-D-glucosaminidase from *Aspergillus fumigatus* S-26.*Protein ExprPuri. 45*, 125–131.

[98] Kim, S. Y., Shon, D. H., & Lee, K. H. (1998). Purification and characteristics of two types of chitosanases from *Aspergillus fumigatus* KH-94. *J Microbiol Biotechnol. 8*, 568–574.

[99] Ike, M., Ko, Y., & Yokoyama, K., et al.(2007). Cellobiohydrolase I (Cel7A) from *Trichoderma reesei* has chitosanase activity. *J Mol Cat B: Enzymatic. 47*,159–163.

[100] Lee,D.X. (2008). The mechanism of chitosan hydrolysis by commercial lipase.*phD Dissertation.*Jiangnan University.Wuxi.

[101] Henrissat B, & Bairoch A. (1996). Updating the sequence-based classifica tion of glycosyl hydrolases. *Biochem J. 316*,695–696.

[102] Henrissat B, & Davies G.(1997). Structural and sequence-based classification of glycoside hydrolases. *Current Opinion in Structural Biology. 7*,637-644.

[103] Honda, Y., Fukamizo, T., & Boucher, I., et al.(1997).Substrate binding to the inactive mutants of *Streptomyces sp.* N174 chitosanase: indirect evaluation from the thermal unfolding experiments. *FEBS Letters.411*, 346-350.

[104] Katsumi, T., Harvey, M.E.L., & Tremblay, H.,et al.(2005). Role of acidic amino acid residues in chitooligosaccharide-binding to *Streptomyces sp.* N174 chitosanase. *Biochem Biophys Res Commun. 338*, 1839-1844.

[105] Fukamizo, T., Juffer, A.H., & Vogel, H.J.,et al.(2000).Theoretical calculation of pKa reveals an important role of Arg205 in the activity and stability of *Streptomyces sp.* N174 chitosanase. *J Biol Chem. 275*, 25633-25640.

[106] Fukamizo, T., Yoshikawa, T., & Katsumi, T.,et al.(2005). Substrate-binding mode of bacterial chitosanases. *J Appl Glycosci. 52*, 183-189.

[107] Saito, J., Kita, A., & Yiguchi, H., et al.(1999).Crystal structure of chitosanase from *Bacillus circulans* MH-K1 at 1.6-angstrom resolution and its substrate recognition mechanism. *J Biol Chem.274*, 30818-30825.

[108] Shimono, K., Shigeru. K., & Tsuchiya, A., et al. (2002). Two glutamic acids in chitosanase A from *Matsuebacter chitosanotabidus* 3001 are the catalytically important residues. *J. Biochem.131*, 87-96.

[109] Blake, C. C., Koenig, D. F., & Mair, G. A., et al.(1965). Structure of hen egg-white lysozyme. A three-dimensional Fourier synthesis at 2 Å resolution. *Nature.206*, 757–761.

[110] Johnson, L. N. & Phillips, D. C. (1965). Structure of some crystalline lysozyme-inhibitor complexes determined by X-ray analysis at 6 Åresolution. *Nature. 206*, 761–763.

[111] Ford, L. O., Johnson, L. N., & Machin, P. A.,et al. (1974). Crystal structure of a lysozyme– tetrasacharide lactone complex. *J Molecular Biol.88*, 349–371.

[112] Siddiqui, K. S., Azhar, M. J., & Rashid, M. H., et al.(1997).Stability and identification of active-site residues of carboxymethylcellulases from *Aspergillus niger* and Cellulomonas biazotea. *Folia Microbiol. 42*, 312–318.

[113] Shoseyov, O., Shani, Z., & Levy, I. (2006).Carbohydrate binding modules: biochemical properties and novel applications. *Microbiol Mol Biol Rev. 70(2)*,283-295.

[114] Bourne, Y., & Henrissat, B.(2001). Glycoside hydrolases and glycosyl transferases: families and functional modules. *Curr Opin Struct Biol.11(5)*,593-600.

[115] Hoare, D.G., & Koshland, D. E. (1967). A method for the quantitative modification and estimation of carboxylic acid groups in proteins. *J Biol Chem.242*, 2447–2453.

[116] Kotsira, V. P., & Clonis, Y. D. (1998). Chemical Modification of Barley Root Oxalate Oxidase Shows the Presence of a Lysine, a Carboxylate, and Disulfides, Essential for Enzyme Activity. *Archives Biochem Biophys. 356*, 117-126.

[117] Siddiqui, K. S., Saqib, A. A. N., & Rashid, M. H., et al.(2000). Carboxyl group modification significantly altered the kinetic properties of purified carboxymethylcellulase from *Aspergillus niger*. *Enzyme Microbial Technol.27*,467–474.

[118] Silva, M. C. P., Terra, W. R., & Ferreira, C. (2004).The role of carboxyl, guanidine and imidazole groups in catalysis by a midgut trehalase purified from an insect larvae. *Insect Biochem Molecular Biol.34*, 1089–1099.

[119] Xia, C., Meyer, D., & Chen, H., et al.(1993). Chemical modification of GSH transferase P1-1 confirms the presence of Arg-13, Lys-44 and one carboxylate group in the GSH-binding domain of the active site. *Biochem J. 293*, 357–362.

[120] Spanda, T. F., & Witkop, B.(1967). Determination of the typtophan content of protein with N-bromosuccinimide. *Methods in Enzymology. 11*,496-506.

[121] Konishi,Y.K., Aoki, T., & Satoh ,N.,et al. (2004).Chemical modification of a catalytic antibody that accelerates insertion of a metal ion into porphyrin: essential amino acid residues for the catalytic activity. *J Molecular Catalysis B: Enzymatic. 31*, 9–17.

[122] Enoch, H.G., & Strittmatter P.(1978). role of tyrosyl and arginyl residuals in rat liver microsomal stearylcoenzyme Adesaturase. *Biochemistry.17*, 4927-4932.

[123] Riordan, J. F., Wacker, W. E., & Vallee, B.L.(1954). N-Acetylimidazole: a reagent for determination of "free" tyrosyl residues of protein. *Biochemistry .4 (9)*, 1758-1765.

[124] Liu,P.,Xia,W.X., & Liu,J. (2008) .The role of carboxyl groups on the chitosanase and CMCase activity of a bifunctional enzyme purified from a commercial cellulase with EDC modification .*Biochem Engin. J. 41*, 142–148.

[125] Duan,W.K., Zheng, C.C., & Zhou, X.Y.,et al. (2007). Culture medium optimization for producing chitosanase by *Trichoderma viride*. *J.Zhejiang.Univer.Technol.35 (1)*,41-45.

[126] Mandels, M., & Reese, E.T.(1957).Induction of cellulose in *Trichoderma viride* as influenced by carbon sources and metals.*J.Bacteriol.73*, 269-278.

[127] Liu, B.D.(2004). Clone and expression of cellulases from *Trichoderma viride.PhD Dissertation*. Harbin Institute of Technology, Biological Engineering Department.Harbin,China.

[128] Shin, C., & Manley, J. L.(2004). Cell signaling and the control of PRE-mRNA splicing. *Nature reviews:Mol.Cell. Biol. 5* ,727-738.

[129] Early, P., Rogers, J., & Davis, M., et al.(1980).Two mRNAs can be produced from a single immunoglobulin μ gene by alternative RNA processing pathways. *Cell. 20*,313-319.

[130] Rosenfeld, M. G., Lin, C. R., & Amara, S. G., et al.(1982)Calcitonin mRNA polymorphism: peptide switching associated with alternative RNA splicing events. *Proc Natl Acad Sci US A.79(6)*, 1717-1721.

[131] ander, E.S., Linton, L.M., & Birren, B.,et al. (2001). Initial sequencing and analysis of the human genome. *Nature. 409*, 860–921.

[132] Modrek, B. & Lee, C.(2002). A genomic view of alternative splicing. *Nat. Genet. 30*,13–19.

[133] Faustino,N,A., &Cooper,T.A. (2003).Pre-mRNA splicing and human disease. *Genes Dev. 17*, 419-437.

[134] Sims, P. F. G., Soares-Felipe, M.S. , &Wang, Q., et al.(1994). Differential expression of multiple exo-cellobiohydrolase I-like genes in the lignin-

degrading fungus Phanerochaete chrysosporium.*Mol. Microbiol. 12,*209–216.

[135] Birch, P. R. J., Sims, P. F. G., & Broda, P.(1995). Substrate-Dependent Differential Splicing of Introns in the Regions Encoding the Cellulose Binding Domains of Two Exocellobiohydrolase I-Like Genes in Phanerochaete chrysosporium. *Applied Environ. Microbial.16(10),* 3741-3744.

[136] Baba, Y., Shimonaka, A., &Koga, J.,et al.(2005).Alternative Splicing Produces Two Endoglucanases with One or Two Carbohydrate-Binding Modules in Mucor circinelloides . *J. Bacteriol.187(9),*3045-3051.

[137] Costanzo, S., Ospina-Giraldo, M.D., & Deahl, K.L.,et al.(2007).Alternate intron processing of family 5 endoglucanase transcripts from the genus Phytophthora. *Curr Genet.52,*115-123.

[138] Gusakov, A.V., Sinitsyn, A. P., & Salanovich, T.N., et al. (2005). Purification, cloning and characterisation of two forms of thermostable and highly active cellobiohydrolaseI (Cel7A) produced by the industrial strain of Chrysosporium lucknowense. *Enzyme.Microbial Technol.36,* 57-69.

[139] **Curach, N.C., Te'O, V.S.J., & Gibbs, M.D.**,et al. (2004).Isolation, characterization and expression of the hex1 gene from Trichoderma reesei. *Gene. 331,*133-140.

[140] Yamada T.,Hiramatsu, S., &Songsri, P.,et al. (1997). Alternative Expression of a Chitosanase Gene Produces Two Different Proteins in Cells Infected with Chlorella Virus CVK2. *Virology. 230,* 361-368.

[141] An, X.P., Lu, J., &Huang, J.D., et al. (2007). Rapid Assembly of Multiple-Exon cDNA Directly from Genomic DNA. *PLoS ONE, ,2(11),* e1179. doi:10.1371/ journal.pone. 0001179.

[142] Liu,P.(2009).Study on the structure and function of bifuntional cellulase-chitosanase from Trichoderma.viride. *PhD Dissertation.* Jiangnan University,School of food science and technology,Wuxi,China.

INDEX